Cyril Norman Hinshelwood

1897	Born, London
1916–18	War research (explosives)
1919	Entered Balliol College
1920	Awarded Oxford degree with distinction
1920	Fellow, Balliol College
1921	Fellow, Trinity College
1926	Research in unimolecular reactions: energy of activation can be overcome, in part, by utilization of internal energy of molecules
1926	Books: *Thermodynamics for Students of Chemistry* and *Kinetics of Chemical Change in Gaseous Systems*
1927	Research on chain reactions, explosions
1929	Elected Fellow of the Royal Society (at age 31)
1934	Book: *The Reaction Between Hydrogen and Oxygen* (with A. T. Williamson)
1937	Fellow, Exeter College, as Dr Lee's Professor of Chemistry
1940	Book: *Kinetics of Chemical Change*
1942	Awarded Davy Medal of the Royal Society
1946	Book: *Chemical Kinetics of the Bacterial Cell;* Bakerian Lecture on the Hydrogen-Oxygen Reaction
1946–8	President of the Chemical Society
1947	Awarded Royal Medal
1948	Knight Bachelor
1951	Book: *The Structure of Physical Chemistry*
1955–60	President of the Royal Society
1956	Awarded the Nobel Prize in chemistry (with N. N. Semenov)
1960	Awarded Order of Merit; also Leverhulme Medal
1961–3	President of the Faraday Society
1962	Awarded Copley Medal
1966	Book: *Growth, Function and Regulation in Bacterial Cells* (with A.C.R. Dean)
1967	Died, London

Chemical dynamics via molecular beam and laser techniques

THE HINSHELWOOD LECTURES OXFORD, 1980

Richard B. Bernstein

Higgins Professor of Natural Science,
Columbia University in the City of New York, and
Senior Vice President, Occidental Research Corp., California.

Clarendon Press • Oxford
Oxford University Press • New York
1982

Oxford University Press, Walton Street, Oxford OX2 6DP

London Glasgow New York Toronto
Delhi Bombay Calcutta Madras Karachi
Kuala Lumpur Singapore Hong Kong Tokyo
Nairobi Dar es Salaam Cape Town
Melbourne Wellington

and associate companies in
Beirut Berlin Ibadan Mexico City

© Oxford University Press 1982

Published in the United States by
Oxford University Press, New York

Library of Congress Cataloging in Publication Data

Bernstein, Richard Barry, 1923–
  Chemical dynamics via molecular beam and
laser techniques.

  (Hinshelwood lectures ; 1980)
  Includes index.
  1. Chemical reaction, Rate of.   2. Lasers in
chemistry.   3. Molecular beams.   I. Title.
II. Series.
QD502.B48   541.3'9   82-3420
ISBN 0-19-855154-1   AACR2
ISBN 0-19-855169-X (pbk.)

British Library Cataloguing in Publication Data

Bernstein, Richard B.
  Chemical dynamics via molecular beam and laser
techniques.
  1. Excited state chemistry
  I. Title
  541.2'8   QD461.5

ISBN 0-19-855154-1
ISBN 0-19-855169-X Pbk

Printing (last digit): 9 8 7 6 5 4 3 2 1

Printed in the United States of America

Foreword

After the death of Professor C. A. Coulson in January 1974, Oxford University was for some years without a professor of Theoretical Chemistry. The IBM Company generously agreed to support a Visiting Professorship for three years and we were glad to welcome Professors R. A. Marcus (Illinois), R. G. Gordon (Harvard) and B. Widom (Cornell). The first two were elected to Fellowships at University College, and the third to one at St. Catherine's.

This scheme came to an end when Professor N. H. March was appointed to the newly-named Coulson Professorship of Theoretical Chemistry in 1977 but it had proved so valuable that we resolved to continue it in another form, and to broaden it to cover the whole field of physical and theoretical chemistry. We therefore instituted a Lectureship which we named after my distinguished predecessor, Sir Cyril Hinshelwood, and which was again coupled with a Visiting Fellowship at St Catherine's College. We were extremely grateful to the Goldsmiths' Company, to whom Hinshelwood had left the residual part of his estate for charitable purposes, for agreeing to support this endeavour. Our first three Lecturers were Professors R. S. Berry (Chicago) and R. B. Bernstein (Columbia) and Dr J. A. Barker (IBM). The course of lectures that each has given has been a most valuable addition to the education of all of us in this Department, and of the many visitors they attracted from elsewhere.

To Professor Bernstein we are particularly grateful for writing up his course as a book, which will spread to an even wider audience the superb course of lectures he gave us in Trinity Term, 1980, on a subject to which he has contributed so much.

Oxford, 1981. J. S. Rowlinson

Preface

I am deeply honoured to have been selected as the Hinshelwood Lecturer at Oxford, for Trinity Term 1980. This is especially significant for me because my own research career has been strongly influenced by the seminal ideas of Sir Cyril Hinshelwood in the field of gas-phase chemical kinetics. In fact, my interest in this subject, which has so dominated my research activities over the past three and a half decades, arose specifically from the study of Hinshelwood's 1940 book, *Kinetics of Chemical Change,* while a graduate student in chemistry at Columbia University during the war years.

I was especially intrigued by Hinshelwood's concept that the activation energy barrier for chemical reaction can be overcome, at least in part, by the utilization of the *internal* energy of the reacting molecules as well as the collision energy derived from relative *translational* motion. Hinshelwood's ideas formed the basis for much research of great current interest in the field of molecular reaction dynamics, as will be seen later on.

The molecular beam technique is now widely used to control the relative velocity of colliding reagents, as is the laser to excite reactants to selected internal states of excitation. It is becoming possible to decide, experimentally, upon the relative efficacy of the reagents' electronic, vibrational, rotational, or mutual translational energy upon their reactivity, and compare with theory. This is one of the major themes of modern chemical dynamics.

The goal of these Hinshelwood Lectures is to bring to the advanced undergraduate student in chemistry a bit of the flavour of current research in the dynamics of elementary gaseous reactions, especially the contributions to this field made by molecular beam and laser techniques. Obviously, in such a short course of lectures, it is difficult to ensure a balanced, or even systematic, presentation of the subject matter. I must therefore apologize in advance for many omissions (not always inadvertent!) of much relevant material, inadequate citations to the original literature, and the somewhat subjective emphasis throughout (unacceptable in a research monograph, I hope tolerable for these 'lecture notes').

In preparation for these Hinshelwood Lectures, I found it of interest to assemble a brief, chronological outline of the distinguished scientific career of Sir Cyril Hinshelwood, identifying some of the significant activities and milestones in his professional life (see p. ii). (Omitted are references to his many accomplishments in the field of arts and humanities: he was a man of extraordinary breadth, clearly the personification of an 'Oxford scholar.') The picture which emerges is that of a truly great figure in science, one whose original research contributions shaped the field of chemical kinetics, earning for Hinshelwood a permanent place in the history of chemistry. It is to his memory that these Lectures are dedicated.

I wish to express my gratitude to Professor John S. Rowlinson, Dr E. Brian Smith, and the other members of the Physical Chemistry Laboratory for making my stay here possible, and fruitful! I am also deeply appreciative of the award of the Christensen Fellowship at St. Catherine's College. It was a privilege to make the acquaintance of the Master, Lord Bullock, and many of St. Catherine's interesting Fellows, from whom I learned much about Snow's 'two cultures,' and how to bridge them.

Richard B. Bernstein
Oxford, June 1980

Note added in proof

The present manuscript has been updated somewhat from the 'lecture notes' of June, 1980, but there is still no pretence of complete literature coverage. However, I have made an attempt to reference the key review articles and monographs published through 1981. I wish to thank my most recent coworkers (F. J. Aoiz, L. M. Casson, K. K. Chakravorty, D. A. Lichtin, M. M. Oprysko, G. B. Spector and D. W. Squire) for their constructive comments on the penultimate draft of this book.

The writer wishes to acknowledge the longtime financial support of his research by the U.S. National Science Foundation. Of course, the author's *principal* acknowledgment is to his family for their continuous moral support (consisting of both encouragement and sacrifice) provided over many years: to parents Simon and Stella, wife Norma, and children Neil, Minda, Beth, and Julie.

One last item may be mentioned as the book goes to press. Since completing the final draft at Columbia University in the City of New York (winter 1981), the writer has undergone a personal transition (of small prior probability, best described by the sudden approximation), namely, from academia to the corporate research world. He is now the Senior Vice President of Occidental Research Corporation in California. Shades of C. P. Snow—are there three cultures?

Richard B. Bernstein
Irvine, February 1982

Contents

1 Introduction: dynamics of elementary chemical reactions

What do we mean by 'chemical dynamics'?

Let us take the area of chemical dynamics to be that part of the field of chemical kinetics that is concerned with the *microscopic, molecular dynamic* behaviour of reacting systems, i.e., with the intra- and intermolecular motions that characterise the elementary act of chemical reaction. We shall exclude the vast and important area dealing with the phenomenological, *macroscopic* description of rate processes for 'bulk' systems. For the purpose of these lectures we shall further limit our scope to the molecular dynamics of systems of atoms and small molecules in the dilute gas phase. This despite the intense current interest in dynamic processes occurring in the condensed phase (e.g., in liquid solutions and in the solid state).

The reasons for this restriction are twofold: first, because it seems preferable to deal with systems sufficiently simple to 'understand' (i.e., describe), and second, because the main emphasis of the present lectures is upon the contribution to chemical dynamics made by the *molecular beam* and *laser* techniques. These are ideally suited to the study of isolated molecules and their mutual interactions. They have also been applied, respectively, to gas-surface systems and to solutes in solution, but these aspects will not be considered in this brief course of lectures.

Perhaps a useful way to characterise a subject is to consider some of the *questions* which lie in its domain. With this aim let us list a number of such questions, central to the field of chemical dynamics in its present state. (Please note, however, that no warranty is offered to the effect that their answers can be found in subsequent pages!)

How do molecules react, i.e., how do 'reagent' molecules come together, 'collide,' 'scatter,' often exchange energy with one another, sometimes break chemical bonds and make new ones, and then separate into product molecules?

How can we distinguish between a reaction which occurs 'directly,' i.e., within an interaction time characteristic of a single molecular vibration period (e.g., $\lesssim 0.1$ ps), and one which proceeds via a relatively stable intermediate complex that lives longer than, say, a typical molecular rotation period (e.g., $\gtrsim 1$ ps)?

What principles determine the course of an overall bimolecular reactive collision, i.e., the probability of formation of a transitory, or a long-lived, intermediate, and its 'unimolecular' decay into products? What factors govern the formation of neutral excited state or ionic products?

What is the origin of the proclivity for the formation of highly excited states of nascent product molecules from fast, exoergic reactions; and of the enhanced reactivity of laser-excited reagents in reactions which must overcome activation energy barriers?

What is the microscopic basis of the termolecular atomic or radical recombination process, and of collision-induced dissociation (or ionization)?

What governs the probability of collisional excitation of internal energy states of molecules, and of the collisional relaxation of disequilibrium populations of excited states to Boltzmann equilibrium distributions? Indeed, how can one characterise *dis*equilibrium internal state distributions, to which the concept of temperature is inapplicable?

How does the lifetime of a vibrationally excited molecule depend upon its energy content in excess of that required to dissociate, isomerize, or ionize? What governs the branching ratios for its decay to yield various alternative product sets (neutral and/or ionic)? Can we selectively excite a specific bond in a molecule and thereby influence the path of the unimolecular decay or enhance bimolecular reactivity at a localised site?

What is the basis of the infrared multiphoton dissociation phenomenon? How is it possible for a powerful infrared laser to excite a selected vibrational mode of a specific isotopic variant of a polyatomic molecule, such as SF_6, and pump successive infrared photons into the same molecule until its total energy exceeds the threshold for bond rupture, e.g.,

$$SF_6 \xrightarrow[(CO_2 \text{ laser})]{nh\nu} SF_6^\dagger \rightarrow SF_5 + F$$

a process requiring $n \gtrsim 35$ photons, in a few tens of ns? How about the analogous process in the visible and ultraviolet, namely, resonance-enhanced multiphoton ionization and fragmentation of polyatomic molecules, e.g.,

$$C_6H_6 \xrightarrow[(\text{dye laser})]{nh\nu_1} C_6H_6^* \xrightarrow{mh\nu_2} \begin{cases} C_6H_6^+ + e^- \\ C_4H_4^+ + C_2H_2 + e^- \\ \cdots \\ C^+ + \cdots + \\ \text{etc.} \end{cases}$$

where $n = 1$ or 2 and m ranges from 1 up to about 10 (depending upon the photon flux), all happening within a few ns?

When a molecule absorbs a single (energetic) photon and photoionization occurs (with concurrent ion fragmentation), what determines the *internal* energy states of the ions thus formed (determined by their coincidence with photoelectrons of given kinetic energy)? What about the neutral fragments from single photon-induced photodissociation? How does photofragmentation ('translational') spectroscopy yield information on the internal state distribution in the photofragments?

Returning to a more basic level, we must firm up a few concepts and definitions. What do we mean by the 'cross section' for elastic scattering, i.e., the differential and total cross sections? What about the analogous cross sections for state-specific inelastic scattering and those for state-to-state reactive scattering? What are the detailed differential (angular) state-to-state cross sections? How are these cross sections related to bimolecular reaction rate coefficients?

How do the differential (and integral) state-to-state reaction cross sections depend upon the collision energy? Upon the relative orientation of the reagent molecules at the moment of impact? Upon the rotational angular momentum (and polarization thereof) of one or both of the reagents? What determines the polarization of the angular momentum of the products? How important is the constraint of angular momentum conservation (in contrast to that of energy conservation) in controlling the dynamics of reaction?

Perhaps we need to enquire more into the theoretical basis of the observable chemical dynamical behaviour. Can we predict anything at all useful from first principles? Supposing we have available an accurate *ab initio* quantum mechanically computed intermolecular potential energy hypersurface for a simple, three-atom, 'reactive' system (A + BC), in the form of a grand table of the electronic energy $V(R_{AB}, R_{BC}, R_{CA})$ over a wide range of internuclear separations R_{ij}. Assuming further that an analytical, or splined, fit is available to represent this surface and that the separate diatomic potentials are all consistent with experiment (e.g., predictive of all observed vibrational levels as well as the known diatom dissociation energies), how do we proceed to deal with the chemical dynamics?

How can we effect a computer simulation of the complicated intra- and intermolecular motions which somehow (occasionally) transform a set of 'reagents' into 'products,' or which fail (most of the time) to do so? Consider first the question of how many types of collisional processes are involved, even, for simplicity, assuming ground state reagents. At low translational energies, only elastic collisions can occur, but as we increase the collision energy, first rotational (j) and then vibrational (v) transitions can be excited; ultimately, reactive processes are allowed. Schematically, as the energy is increased, we must expect a succession of elementary processes:

$$
A + BC \rightarrow
\begin{cases}
A + BC & \text{elastic scattering} \\
A + BC(j) & \text{rotational excitation} \\
A + BC(j,v) & \text{vibrational (plus rotational) excitation} \\
AB + C & \text{exchange reaction (1)} \\
AC + B & \text{exchange reaction (2)} \\
A + B + C & \text{collision-induced dissociation}
\end{cases}
$$

(Note that electronic excitations and ionization processes have been excluded; for simplicity, we assume that a single, lowest-lying, adiabatic potential surface is all that is involved in the above-listed processes.)

Can we describe the dynamics by the exclusive use of classical mechanics, i.e., by 'following' a great many classical trajectories, sampling over the various 'initial conditions,' and appropriately averaging the end results? We know that we must somehow develop 'quantisation' of the product's states, for the observations to be simulated are not merely 'continuous' angular distributions but cross sections for populating specific internal states. What tricks can be used to recover from these computations the essence of the observable elastic, inelastic, and reactive scattering of even the simplest atom-molecule systems? Or is it going to be necessary to solve (numerically, at vast expense), the so-called multichannel, three-dimensional (3-D) quantal scattering problem? (The solution of the full Schrödinger equation for the system usually involves the solution of a great many coupled integro-differential equations. It is a *tour de force* operation justifiable only as a 'benchmark' calculation.) Can semiclassical methods be used to obviate much of the labour of the quantum mechanical solution? What about shortcuts, such as a solution in one dimension (1-D), or two (2-D)?

How about turning the problem around and asking what we can learn about the potential energy surface from observable scattering behaviour, when combined with spectroscopic data? How much about the potential is really determined by the various kinds of observations at different levels of detail, e.g., differential cross sections for elastic, inelastic, and reactive scattering over a wide range of collision energies? Are these observations governed mainly by the topology of the surface, insensitive to details, or do they reveal quantitative aspects of its topography?

Under what conditions can we 'invert' experimental data on elastic (and inelastic) scattering to yield the interaction potential? What about making use of bulk thermophysical properties for this purpose (e.g., virial and transport coefficients, acoustic absorption and dispersion)?

How can we make use of spectroscopically derived (e.g., via molecular beam electric resonance experiments) structures of van der Waals molecules, say A $\cdots$ BC, to deduce the potential energy surface that is appropriate for describing the elastic and inelastic scattering of BC by A?

How much information (and how much redundancy?) is there in a set of state-to-state cross sections? How can information theory be used to 'squeeze' the most physical significance out of the plethora of detailed experimental (or computer-simulated) state-to-state data, and how 'informative' are the various kinds of experiments in the field of chemical dynamics?

What is the goal of all this enquiry? Is it perhaps to gain the ability to describe—or better, to calculate—from first principles, the observable dynamic behaviour of chemically interesting systems? Does this represent 'understanding' of chemical dynamics?*

More to the point of these lectures, what are the kinds of *observations* we can make that bear upon these questions? How can we decide upon, then measure (and, in due course, account for, theoretically) the 'right quantities,' e.g., how can we measure the products' angular and recoil velocity distribution for an elementary chemical reaction carried out under controlled conditions (i.e., known relative velocity and reagent internal state distributions)? What about the ascertainment of relative populations of products' internal states? How do we measure differential and integral reaction cross sections for 'state-to-state' processes, in general? How can we determine the dependence of the reaction probability and the products' angular and energy distributions upon the reagents' internal energies and their mutual collision energy?

At a more practical level, how can we form a collimated beam of molecules with a given velocity, so as to carry out collision experiments at selected values of the relative velocity, say v_r. (A given value of v_r corresponds to a given de Broglie wavelength $\lambda = h/\mu v_r$, where μ is the reduced mass of the colliding pair and h is Planck's constant; 'monochromatic' scattering experiments are surely to be desired.) How can we accelerate molecules to achieve superthermal beams with translational energies above about 1 eV (up to several kV)?

How can we detect neutral molecules at low number densities? How can we measure their velocity distribution? How can we ascertain the electronic, vibrational, and rotational state distribution of molecules, especially nascent products of chemical reactions?

How can we prepare beams of radicals or normally unstable reagent molecules? What about electronically excited atoms and molecules?

What are the various means of preparing beams of state-selected molecules, e.g., rotationally and/or vibrationally selected diatomics or internally excited polyatomics? How can we form polarized beams?

How can we orient molecules in a beam so that we can observe the dependence of reactivity upon relative orientation with respect to an incident co-reagent? What kinds of molecules can be oriented, and what factors govern the degree of precession of the molecules about the orientation axis?

How can we 'couple' lasers with molecular beams? What are the factors limiting our ability to prepare intense state-selected beams of polyatomic molecules? How important is the power density (or intensity) of the laser radiation (in terms of the number of photons m^{-2} s^{-1}, or watts cm^{-2}) relative to the

---

*Many of the above questions have been discussed in some detail in two books co-authored by this writer: *Molecular Reaction Dynamics* (R. D. Levine and R. B. Bernstein. Oxford: Clarendon Press, 1974), hereafter abbreviated MRD or LEV74, and *Atom-Molecule Collision Theory—A Guide for the Experimentalist* (R. B. Bernstein, ed. New York: Plenum Press, 1979), BER79.

fluence, or fluency, (in terms of the *total* number of photons m^{-2} per pulse, or joules cm^{-2} per pulse) in promoting multiphoton 'up-the-ladder' processes? How important is coherence in the photon beam; and intramolecular dephasing? Will multicolor laser experiments (UV + IR, vis + UV, etc.) yield new routes to specific fragmentation products?

Can intense laser fields influence the transition state in a chemical reaction, altering reaction paths as well as cross sections?

What new discoveries will ensue from the successful development of the pulsed, high duty factor, high 'wall-plug efficiency,' free-electron laser, tunable over several wide regions of wavelengths, i.e., regions spanning the range from $\lesssim 1$ Å to $\gtrsim 100$ μm? Isotope separation? Bond-selective chemistry? What next?

What is the *future* of chemical dynamics? To answer this, we should first look at its *past* and *present*—the purpose of these lectures.

2 State-to-state reaction cross sections and rate coefficients: theory and experiment

Let us begin by casting a backward look to the turn of the century; the work of Bodenstein following upon Arrhenius had already established much of the essential phenomenology of gas-phase kinetics. By the time of Hinshelwood's first book in 1926, the gross kinetic behaviour of a reacting gaseous mixture was well characterised experimentally. Although a satisfactory understanding of unimolecular reactions was not yet in hand, the bimolecular process had been correctly interpreted on a microscopic basis, a collision-induced mechanism of reaction. Let us go on to see how much more we may have learned over the past half-century or so.

2.1. Kinetics of elementary bimolecular reactions

For the present purpose, we shall consider a very simple system, for example, an exoergic atom-molecule 'exchange' reaction, such as

$$Cl + HI \rightarrow HCl + I \tag{2.1}$$

carried out at a given temperature T, well studied by J. C. Polanyi and co-workers. Postponing, for the moment, the examination of this particular reaction, let us discuss the general three-atom case, the one jocularly designated the argon−boron carbide reaction:

$$A + BC \rightarrow AB + C \tag{2.2}$$

The initial rate of reaction, say R_0, is normally found to depend upon the product of the number densities of the reagents:

$$R_0 (\equiv \dot{n}_{AB} = \dot{n}_C = -\dot{n}_A = -\dot{n}_{BC}) = k(T)n_A \cdot n_{BC} \tag{2.3}$$

where $k(T)$ is the thermal 'rate constant' for the reaction carried out at the temperature T. For many reactions, k is a strongly rising function of T (Fig. 2.1 left). An Arrhenius plot of $\log_{10} k$ versus $1/T$ is nearly linear (Fig. 2.1 right). The old rigid-sphere 'collision theory,' invoking the so-called line-of-centres condition for reaction, leads to the Arrhenius-like result for the rate constant:

$$k(T) = p\bar{v}\sigma^0 \exp(-E_a/RT) \tag{2.4}$$

where $\bar{v} = (8RT/\pi\mu)^{1/2}$ is the average relative speed of the reagents (of reduced mass μ) and $\sigma^0 \equiv \pi d^2$ is a 'rigid-sphere cross section' (with d the

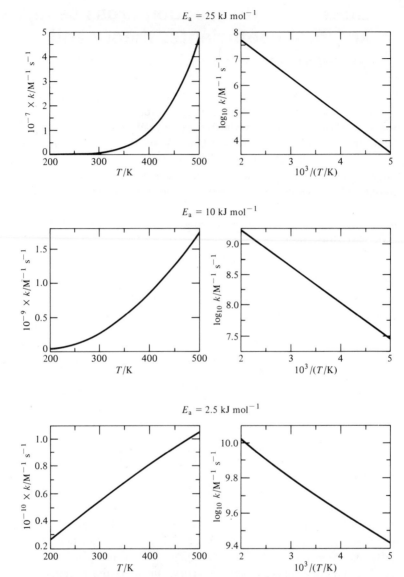

FIG. 2.1. *Left:* temperature dependence of bimolecular rate constants for indicated activation energies. *Right:* corresponding Arrhenius plots. From R. J. LeRoy (unpublished).

'collision diameter'). Here, E_a is the activation energy, $E_a = \frac{1}{2}\mu v_0^2$, where v_0 is the minimum value of the projection of the relative velocity along the line of centres of the colliding spheres required to effect reaction. The 'steric factor' p is the probability that the orientation of the spheres at impact is suitable for reaction to occur.

The graphs shown in Fig. 2.1 correspond to $p = 0.25$, $\mu = 25$ amu, $d = 0.25$ nm, and $E_a = 2.5$, 10, and 25 kJ mol^{-1}. The slight curvature in the Arrhenius plots arising from the $T^{1/2}$ exponential factor is very difficult to detect!

2.2. State-to-state rates and cross sections

Figure 2.2 shows an energy-level diagram depicting the activation energy for the formation of the lowest-energy configuration of some transitory (?) adduct ABC, the 'activated complex.' (Let us not elaborate upon this term here, nor

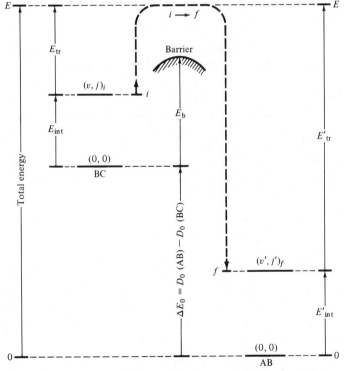

FIG. 2.2. Schematic energy level diagram for the exoergic reaction (with activation barrier) A + BC → AB + C, serving to define terms such as total energy, internal energy, and translational energy of reactants and products. Adapted from R. B. Bernstein and R. D. Levine, BER75.

be sidetracked by transition state theory, which is, in effect, a bypass around the problem of the dynamics of the reactive collision.)

We know from Fig. 2.2, however, that even in the simplest case we must take cognizance of the existence of rotational and vibrational energy states of reagent and product molecules, i.e., the *internal* energy content of the molecules. We must consider the question of the dependence of the rate coefficient upon the initial internal state i (of the reagent) and the final state f (of the product). The overall bimolecular rate coefficient at the temperature T can be expressed as a sum over initial states:

$$k(T) = \sum_i f_i(T) k_i(T) \tag{2.5}$$

where $f_i(T)$ is the fraction of the population of reagents in some initial state (or set of states), designated i, at the temperature T and $k_i(T)$ is the total rate coefficient for all processes leading out of state i to produce various final states f:

$$k_i(T) = \sum_f k_{fi}(T) \tag{2.6}$$

Here $k_{fi} \equiv k(i \rightarrow f)$ is the specific 'state-to-state' rate coefficient for the conversion of reagents in an initial selected state i to products in a final specified state f, at given T.

Returning to Fig. 2.2, we note the designation of the energy of the reagent BC in a particular initial state i, with precollision vibrational and rotational quantum numbers v and j, and of the product AB in a particular final state f, characterised by postcollision quantum number v', j'. The diagram refers to a collision in which the relative translational energy, E_{tr}, plus the internal energy of the reagent, E_{int}, exceeds the 'barrier energy,' E_b. The total energy of the system, E, is defined (for the exoergic reaction of Eqn. 2.2) as shown in Fig. 2.2:

$$E = E'_{tr} + E'_{int} = E_{tr} + E_{int} - \Delta E_0 \tag{2.7a}$$

where $-\Delta E_0$ is the zero-point exoergicity of the reaction. The internal energy includes rotational, vibrational, and electronic energy (the latter not considered explicitly in Fig. 2.2). The total energy is usually measured from the lower of the two zero-point levels, either reagents or products. Thus, for an endoergic reaction ($\Delta E_0 > 0$), the total energy is simply the sum of the translational and reagents' internal energy

$$E = E_{tr} + E_{int} = E'_{tr} + E'_{int} + \Delta E_0 \tag{2.7b}$$

[In the special case of the three-atom system of Eqn. 2.2, $-\Delta E_0$ is simply the difference in bond dissociation energies, $D_0(AB) - D_0(BC)$.]

From the viewpoint of the present lectures, we are concerned with molecular beam measurements of reactive collisions, usually carried out at more or less well-specified relative translational energies determined by the experimental distribution, say $p(v_r)$, of relative speeds $v_r = (2E_{tr}/\mu)^{1/2}$. A given experiment is thus characterised by a certain average collision energy,

$$\overline{E}_{tr} = \int_0^\infty dE_{tr} E_{tr} P(E_{tr}|\overline{E}_{tr}) \tag{2.8}$$

where $P(E_{tr}|\overline{E}_{tr})$ is the probability density function for the relative translational energy at the given $\overline{E}_{tr}$. Note that $P(E_{tr})dE_{tr} = p(v_r)dv_r$. For the given $\overline{E}_{tr}$, the state-to-state rate coefficient $k_{fi}(\overline{E}_{tr})$ is the average, over the distribution of E_{tr}, of the product of the relative speed v_r and the so-called state-to-state reaction cross section σ_{fi}:

$$k_{fi}(\overline{E}_{tr}) = \langle v_r\sigma_{fi}(E_{tr})\rangle_{\overline{E}_{tr}} = (2/\mu)^{1/2} \int_0^\infty dE_{tr} E_{tr}^{1/2} \sigma_{fi}(E_{tr}) P(E_{tr}|\overline{E}_{tr}) \tag{2.9}$$

The total rate constant out of state i is then

$$k_i(\overline{E}_{tr}) = (2/\mu)^{1/2} \int_0^\infty dE_{tr} E_{tr}^{1/2} \sigma_i(E_{tr}) P(E_{tr}|\overline{E}_{tr}) \tag{2.10}$$

where

$$\sigma_i = \sum_f \sigma_{fi} \tag{2.11}$$

is the reaction cross section out of state i.

In the case of a nonbeam experiment with a Maxwellian distribution of E_{tr} at temperature T, $\overline{E}_{tr} = \tfrac{3}{2}kT$ and

$$P(E_{tr}) = 2(E_{tr}/\pi)^{1/2}(kT)^{-3/2} \exp(-E_{tr}/kT) \tag{2.12}$$

where $k \equiv R/N_{Avog}$ is Boltzmann's constant.

Substituting Eqn. 2.12 into Eqn. 2.10 yields

$$k_i(T) = (\pi\mu)^{-1/2}(2/kT)^{3/2} \int_0^\infty dE_{tr} E_{tr}\sigma_i(E_{tr})\exp(-E_{tr}/kT) \tag{2.13}$$

as shown by M. Eliason and J. O. Hirschfelder.

If somehow we knew $\sigma_i(E_{tr})$, we could calculate $k_i(T)$; then, given the initial state population distribution $f_i(T)$, we could calculate $k(T)$ from Eqn. 2.5. Supposing we assume for want of further information that σ_i is independent of i. Further suppose it has only a simple line-of-centres translational energy dependence for $E_{tr} \geq E_0$:

$$\sigma_i(E_{tr}) = \sigma(E_{tr}) = p\sigma^0(1 - E_0/E_{tr}) \tag{2.14}$$

Substitution of Eqn. 2.14 into Eqn. 2.13 yields for $k(T)$ the familiar Arrhenius-like result, Eqn. 2.4.

Of course, the simplicity of this result in no way justifies the preceding naive suppositions. So let us return to the general case, i.e., Eqn. 2.9 for k_{fi} in terms of σ_{fi} (and Eqn. 2.10 for k_i in terms of σ_i). Obviously, we need information on the state-to-state cross sections $\sigma_{fi}(E_{tr})$, which are, in general, strongly dependent upon i and f. It is convenient for theoretical considerations to compare state-to-state cross sections not at a particular value of E_{tr} bur rather at a given total energy E.

2.3. Implications of microscopic reversibility

The principle of microscopic reversibility requires that, at any value of E, each of the state-to-state transition probabilities for the forward reaction (from states i to f) is equal to those for the reverse reaction (from that same set of states f to the 'original' set i), i.e., $P_{fi}(E) = P_{if}(E)$, as pointed out by J. Ross, J. C. Light, and K. E. Shuler. These transition probabilities are closely related to the state-to-state cross sections via the so-called scattering matrix (S-matrix) elements. Let us see the implications with respect to Reaction 2.2, rewritten to deal with this issue:

$$A + BC(v,j) \underset{\sigma'}{\overset{\sigma}{\rightleftharpoons}} AB(v',j') + C \tag{2.15}$$

where $i \equiv v,j$; $f \equiv v',j'$; and σ and σ' denote $\sigma_{fi}(E)$ and $\sigma_{if}(E)$, respectively. Then it turns out that

$$\sigma_{fi}(E) = (\pi/k)^2 g' P_{fi}(E) \tag{2.16a}$$

and

$$\sigma_{if}(E) = (\pi/k')^2 g P_{if}(E) \tag{2.16b}$$

where k and k' are the so-called wavenumbers ($k = 2\pi/\lambda$), pre- and postcollision: $k = \mu v_r/\hbar$, $k' = \mu'v'_r/\hbar$, i.e., $k^2 = 2\mu E_{tr}/\hbar^2$, $k'^2 = 2\mu'E'_{tr}/\hbar^2$; g and g' are the rotational degeneracies ($g = 2j + 1$, $g' = 2j' + 1$); and reduced masses are $\mu = \mu_{A,BC}$ and $\mu' = \mu_{AB,C}$. Thus, one obtains the ratio

$$\frac{\sigma_{fi}(E)}{\sigma_{if}(E)} \equiv \frac{\sigma(E)}{\sigma'(E)} = \frac{g'\mu'E'_{tr}}{g\mu E_{tr}} \tag{2.17}$$

or, explicitly, for Reaction 2.15:

$$\left[\frac{\sigma(BC_{v,j} \rightarrow AB_{v',j'})}{\sigma'(AB_{v',j'} \rightarrow BC_{v,j})}\right]_E = \frac{m_{AB}m_C}{m_{BC}m_A} \times \frac{E'_{tr}}{E_{tr}} \times \frac{(2j' + 1)}{(2j + 1)} \tag{2.18}$$

From Eqns. 2.7,

$$E_{tr} = E + \Delta E_0 - E_{int}(BC) \tag{2.19a}$$

and

$$E'_{tr} = E - E_{int}(AB) \tag{2.19b}$$

In the special case of constant E, the state-to-state rate coefficients are simply products of cross section and relative speed:

$$k_{fi}(E) = v_r \sigma_{fi}(E) = 2(E_{tr}/\mu)^{1/2} \sigma_{fi}(E) \tag{2.20}$$

and similarly for $k_{if}(E)$, so that we obtain the ratio of forward to reverse state-to-state rates:

$$\frac{k_{fi}(E)}{k_{if}(E)} = \frac{\mu' g'}{\mu g} \left(\frac{E'_{tr}}{E_{tr}} \right)^{1/2} \tag{2.21}$$

Thus we see that if a particular transition has a 'high' cross section or rate coefficient in the forward direction, its inverse will also have a relatively high cross section or rate.

If in the exoergic Reaction 2.2 it is found that, for dynamic reasons, certain excited states of the products (here AB) are preferentially populated, then for the reverse, endoergic reaction (AB + C) carried out at the same total energy, Eqn. 2.21 implies preferential reactivity for these particular states of AB in reaction with C. This 'selective reactivity' is a dynamic rather than an energetic effect since E is constant, as pointed out by J. C. Polanyi, R. B. Bernstein, and co-workers.

Let us consider the implications of the above, i.e., the possibility of 'vibrational enhancement' of molecular reactivity, by examining a class of fast 'chemical laser reactions,' e.g.,

$$F + H_2 \rightarrow HF^\dagger + H \tag{2.22}$$

where † denotes vibrational excitation. To achieve lasing, 'population inversion' must occur; lasing between two rovibrational states of the product molecule (e.g., $HF^\dagger$) requires the inequality

$$N(v',j')/g' > N(v,j)/g \tag{2.23}$$

where the prime denotes the upper state, $g (= 2j + 1)$ is the rotational degeneracy, and N is the number density of the designated states of the product molecules.

Figure 2.3 is a vibrational energy level diagram for the $F + H_2$ chemical laser system, discovered by G. C. Pimentel and co-workers. The exoergicity of

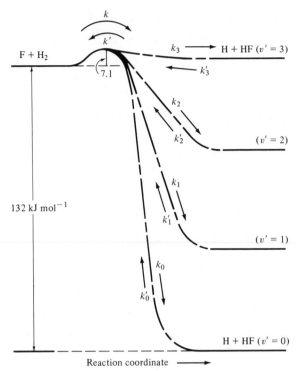

FIG. 2.3. Reaction coordinate and vibrational energy diagram appropriate to the F + H_2 reaction, indicating the various rate constants needed to characterise the state-to-state reaction kinetics.

Reaction 2.22 is sufficient to make accessible (at thermal energies, with $\overline{E}_{tr} \gtrsim$ 7 kJ mol^{-1}) four HF vibrational states ($v' = 0, \ldots, 3$); however, each succes-sive vibrational state is allowed successively fewer rotational states for a given total E. Experiment has revealed that the rate constant k_2, i.e., the rate con-stant for the formation of HF ($v' = 2$), is the largest of the set. Thus, if we carried out the reverse reaction, i.e.,

$$H + HF(v) \rightarrow H_2 + F \qquad (2.24)$$

at the same total energy (which would require the addition of considerable translational energy), HF($v = 2$) would react with H atoms preferentially with respect to the ground state or even the $v = 3$ state of HF. Thus we can expect to have vibrational enhancement of reactivity for the endoergic reaction (Eqn. 2.24).

Many bulk gas-phase experiments are carried out at constant temperature rather than at constant total energy. Sometimes a laser is used to produce a disequilibrium internal state distribution, i.e., rovibrationally excited molecules in the gas. However, in the course of a few dozen collisions, the rotational population has relaxed ('Boltzmannized') to that corresponding to the ambient translational temperature. Since the vibrational relaxation process is usually much slower, the net effect is to have produced a 'metastable' excess population of several particular vibrationally excited states of the molecule in a Boltzmann thermal bath.

Assuming that the rotational distribution of both reagent and product molecules are Boltzmannized (thermalized), detailed balance considerations (as shown by R. D. Levine, J. Manz, and co-workers) lead to the following ratio of rate coefficients:

$$\frac{k(v \rightarrow v' \mid T)}{k'(v' \rightarrow v \mid T)} = \frac{Q'_{\text{rot}}(T)}{Q_{\text{rot}}(T)} \left(\frac{\mu'}{\mu} \right)^{3/2}$$
$$\times \exp[-(E'_{\text{vib}} - E_{\text{vib}})/kT]\exp(-\Delta E_0/kT) \quad (2.25)$$

Here, k denotes the state-to-state rate coefficient at temperature T for the 'forward' reaction, i.e., from reagent vibrational state v to product state v', and k' the corresponding rate coefficient for the inverse reaction from the erstwhile product, now the *reagent*, in vibrational state v' to the original reagent, now the *product*, in state v. The rotational partition functions are standard:

$$Q_{\text{rot}}(T) = \sum_j (2j + 1)\exp(-E_j/kT) \quad (2.26)$$

where E_j is the rotational energy of the original reagent molecule (with an analogous expression for the product). In Eqn. 2.25, E_{vib} and E'_{vib} are the vibrational energies of reagent and product corresponding to the vibrational quantum numbers v and v', respectively. The ratio $(\mu'/\mu)^{3/2}$ reduces to $(m_{\text{AB}}m_\text{C}/m_\text{A}m_{\text{BC}})^{3/2} = Q'_{\text{tr}}/Q_{\text{tr}}$.

Of course, if we carry this to the limit, and assume all *vibrations* to be Boltzmannized, Eqn. 2.25 reduces to the familiar equilibrium constant expression:

$$\frac{k(T)}{k'(T)} = \frac{Q'_{\text{tr}}}{Q_{\text{tr}}} \frac{Q'_{\text{rot}}}{Q_{\text{rot}}} \frac{Q'_{\text{vib}}}{Q_{\text{vib}}} \exp(-\Delta E_0/kT) = K_{\text{eq}}(T) \quad (2.27)$$

2.4. Vibrational enchancement of reactivity

Consider a given exoergic reaction carried out at a specified temperature T, in which the vibrational state of the reagent is selected at some value v (which fixes E_{vib}). From Eqn. 2.25, the dependence of the ratio $k_{v \rightarrow v'}/k'_{v' \rightarrow v}$ upon v' (or E'_{vib}) can be found. Thus,

$$\frac{k_{v \rightarrow v'}(T)}{k'_{v' \rightarrow v}(T)} = \frac{Q'_{\text{rot}}Q'_{\text{tr}}}{Q_{\text{rot}}Q_{\text{tr}}} \times A_v(T)\exp(-E'_{\text{vib}}/kT) \quad (2.28)$$

where

$$A_v(T) \equiv \exp[-(\Delta E_0 - E_{\text{vib}})/kT] \tag{2.29}$$

is independent of v'.

Thus we have

$$\ln(k'/k) = \frac{E'_{\text{vib}}}{kT} - \ln \left(A_v \frac{Q'_{\text{rot}}Q'_{\text{tr}}}{Q_{\text{rot}}Q_{\text{tr}}} \right) \tag{2.30}$$

dropping subscripts on k and k'. Equation 2.30 implies a strong positive exponential dependence of the ratio k'/k upon E'_{vib} (at constant T and E_{vib}).

Thus, even in the *absence* of population inversion in the forward (exoergic) reaction, there is sure to be vibrational enhancement for the inverse (endoergic) reaction, arising trivially from the reduction of the 'net endoergicity' of the latter with increasing E'_{vib}. This is brought out more clearly when we consider the dependence upon E'_{vib} explicitly, from the differential expression obtainable from Eqn. 2.30:

$$\left(\frac{\partial \ln(k'/k)}{\partial E'_{\text{vib}}} \right)_T = 1/kT \tag{2.31}$$

This simple result for the 'vibrational enhancement effect' is valid irrespective of the value of v (or E_{vib}).

A more interesting aspect is the *additional* vibrational enhancement of reactivity (for the inverse reaction) due to the oft-found, dynamically induced population inversion of the forward reaction, i.e., the *increase* in $k_{v \to v'}$ with v' mentioned in connection with the HF laser reaction. It would be desirable to have an estimate of the vibrational enhancement of the *total* rate constant for the endoergic reaction, i.e., $k_{v'}(T) = \Sigma_v k_{v' \to v}(T)$. A simple approximation, due to J. H. Birely and J. L. Lyman, has been used to predict the ratio of the rate coefficients for a given state v' versus the $v' = 0$ state, i.e., $k_{v'}/k_0$. The assumption is made that some given fraction α of the vibrational excitation energy E'_{vib} of the reagent serves to reduce the activation energy E_a of the reaction. Then

$$k_{v'}/k_0 = \exp[-(E_a - \alpha E'_{\text{vib}})/kT] \tag{2.32}$$

so that

$$\left(\frac{\partial \ln k_{v'}}{\partial E'_{\text{vib}}} \right)_T = \alpha/kT \tag{2.33}$$

(cf. Eqn. 2.31).

Studies of several reactions have indicated that α can exceed unity, implying an 'extra effectiveness' of vibrational excitation in promoting the endoergic (inverse) reaction over and above the simple 'reduction in activation energy' implied by Eqn. 2.32.

2.5. Examples of state-to-state reactions

Next, let us look at a few experimental examples of state-to-state reactions.

Figure 2.4 shows relative rate constants for the formation of various vibrational states of barium monohalides from the exoergic reactions of barium

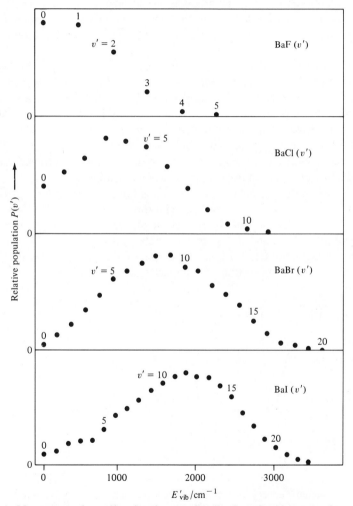

FIG. 2.4. Nascent product vibrational state distributions for the crossed molecular beam reactions of Ba with HF, HCl, HBr, and HI, as measured by laser-induced fluorescence. Adapted from H. W. Cruse, P. J. Dagdigian, and R. N. Zare in FAR73, p. 277.

atoms with hydrogen halides:

$$Ba + XH \rightarrow BaX + H \qquad (2.34)$$

(X ≡ F, Cl, Br, I). The experiments, by R. N. Zare and co-workers, were carried out under 'single-collision' conditions (Ba beam, crossed HX beam target). Laser-induced fluorescence was used to measure the relative populations of the various vibrational levels of the nascent BaX molecules (formed in the ground electronic state). The vibrational population inversions are immediately recognized in the pattern of the products' vibrational populations. These experiments did not reveal detailed information on the products' rotational distributions, however, because of experimental limitations (involving resolution and relaxation; see Chapter 4).

Figure 2.5 is a schematic energy level diagram appropriate to the Cl + HI reaction (Eqn. 2.1), studied extensively by J. C. Polanyi and co-workers using infrared chemiluminescence detection. By working with a jet plus fast-flow system under very low pressure conditions (with suitable cryogenics), it was possible to measure complete rotational-vibrational state distributions in the nascent, vibrationally excited HCl molecules. Shown in the insert of Fig. 2.5 are the relative rates of formation of HCl in the specified vibrational states. A fuller display of more extensive results is presented in Fig. 2.6, i.e., the complete rotational state population distributions for the nascent HCl product in various excited (i.e., infrared-emitting) vibrational states ($v' \geqslant 1$). Results are shown for two different values of $\overline{E}_{tr}$. At the higher energy, a larger number of rotational and vibrational states are accessible (and are found to be populated) and the average internal energy of the product molecules is increased, though the *total* reaction cross section has decreased.

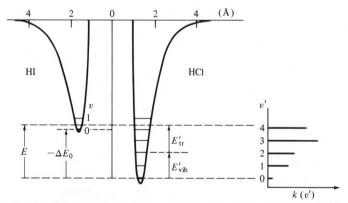

FIG. 2.5. Schematic energy level diagram appropriate to the Cl + HI reaction, showing relative rates of formation of HCl(v') product vibrational states as determined from chemiluminescence experiments by D. H. Maylotte, J. C. Polanyi, and K. B. Woodall, *J. Chem. Phys.* **57**, 1547 (1972). See also MRD.

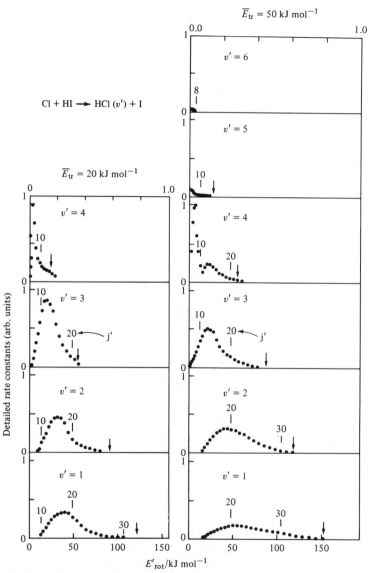

FIG. 2.6. Dependence of product's nascent vibrational and rotational state distribution upon reagents' relative translational energy, for the Cl + HI reaction. Plotted are the relative rate constants for the formation of the indicated rotational and vibrational states versus the rotational energy of the HCl for two different specified values of the average collision energy. Arrows indicate nominal energy conservation limits. Adapted from L. T. Cowley, D. S. Horne, and J. C. Polanyi, *Chem. Phys. Lett.* **12**, 144 (1971); details therein.

Similar detailed rate constants for vibrational-rotational product state formation have been measured for other reactions, and from them it is possible to predict distributions of products' translational energies, say $P(E'_{tr})$, since the total available energy E is (approximately) known. (For the strongly exoergic reactions studied, with small activation barriers, the available energy is mainly due to the exoergicity plus the reagents' relative translational energy $\overline{E}_{tr}$.) So-called continuum detailed rate constants, proportional to $P(E'_{tr})$, i.e., translational distributions calculated from internal excitation data for several reactions, are plotted in Fig. 2.7. The vibrational structure in these translational 'spectra' are evident, and one can immediately recognize the population inversions, responsible for lasing action.

2.6. Potential energy surfaces and models for reaction dynamic behaviour

How can we explain the observed 'energy partitioning,' i.e., the way in which the total available energy is disposed among the various internal degrees of freedom of the product molecules and their relative translational motion? What is the dynamic basis of the vibrational population inversion and the disequilibrium distribution of rotational states? There are different levels of 'explaining' experimental data, of course, ranging from simple models to exact theories. For the present problem, where the probability of a tidy, analytical-theoretical description is small, the most one can reasonably expect is a computational description. One would like to start with an 'accurate' *ab initio* potential energy surface (such as those computed by H. F. Schaefer, W. A. Goddard, C. E. Dykstra, T. H. Dunning, and others) and perform 'exact' quantal scattering calculations (such as those of G. C. Schatz and of A. Kuppermann, R. E. Wyatt, and co-workers A. B. Elkowitz and M. J. Redmon) of the detailed state-to-state cross sections and rate coefficients. The degree to which the computational results simulate the data is then an overall indication of the 'accuracy' of the surface and the 'exactness' of the scattering calculations. In favourable cases, it could then be said that the experimental results have been 'explained'; but perhaps not yet 'understood.'

To gain understanding of the dynamics, some sort of a model, though approximate, may be more helpful. Often we can be satisfied with a qualitative picture, although the more quantitative the better.

A very simple way to account for product vibrational excitation (or the lack thereof) in an exoergic exchange reaction (Eqn. 2.2) and in the inverse endoergic reaction, i.e., the 'vibrational enhancement' effect, was proposed by H. Eyring, G. E. Kimball, and J. O. Hirschfelder in connection with their early work on potential surfaces and rudimentary classical mechanical trajectory calculations. We consider the behaviour of typical 'system trajectories' on two extremes (topologically) of potential surfaces, shown in Fig. 2.8, one with an

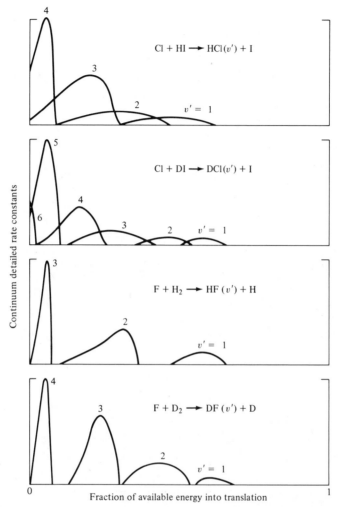

FIG. 2.7. Plots of the products' relative translational energy distributions for the indicated reactions based on chemiluminescence measurements of vibrational and rotational distributions. Adapted from K. G. Anlauf, P. E. Charters, D. S. Horne, R. G. McDonald, D. H. Maylotte, J. C. Polanyi, W. J. Skrlac, D. C. Tardy, and K. B. Woodall, *J. Chem. Phys.* **53**, 4091 (1970).

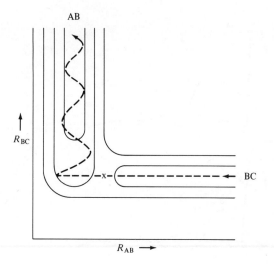

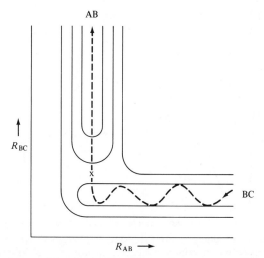

FIG. 2.8. Stylized potential surfaces for collinear, exoergic A + BC reaction dynamics. For the early-barrier case (upper), the reagents' relative translational energy is efficiently converted to product vibration; for the late barrier case (lower), reagent's vibration is efficiently utilized to overcome the barrier and the exoergicity is released in product translation. Adapted from J. O. Hirschfelder, *Intl. J. Quantum Chem. Symp.* **3**, 17 (1969).

'early barrier,' the other a 'later barrier,' intended to represent the collinear A + BC reaction. (The surfaces are highly stylized for simplicity.)

For the early barrier case, the reagents' relative translation is efficiently converted to product's (AB) vibration. For the late barrier, vibrational energy of the molecular reagent (BC) is required in order to 'turn the corner' and overcome the barrier, and then vibrationally 'cold' product is obtained. (For the inverse of these processes, i.e., the 'endo' reactions on the same stylized surfaces, play the tape backwards!)

Figure 2.9 shows some more realistic examples of this behaviour, based on numerical calculations of 'realistic' potential surfaces by J. C. Polanyi and W. H. Wong.

In view of the paucity of *ab initio* potential surfaces of so-called chemical

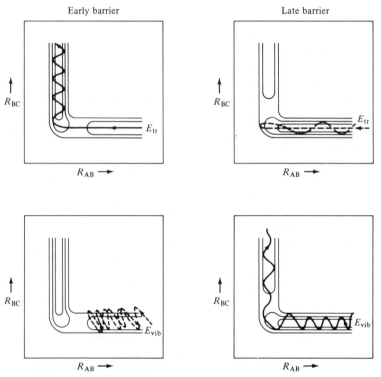

FIG. 2.9. Classically calculated trajectories on stylized potential surfaces for collinear A + BC reactions, showing dynamic effect of early versus late barrier. For high relative translational energies, reaction occurs for early barrier (top left) but not for late barrier (top right). For high vibrational energy of the BC reagent, reaction fails for early barrier (bottom left) but occurs for late barrier, i.e., the reagent's vibration assists in surmounting the (late) barrier. Adapted from J. C. Polanyi and W. H. Wong, *J. Chem. Phys.* **51**, 1439 (1969). See also MRD.

accuracy and the difficulty of the numerical solution of the full quantal (3-D) scattering problem for reactive systems of experimental interest (even for three-atom reactions), a compromise has been reached, involving the use of the so-called quasiclassical trajectory method, which goes something like the following.

First one constructs a trial semiempirical potential energy surface $V(R_{AB}, R_{BC}, R_{CA})$ for the system, based upon the LEPS, or diatomic-in-molecules (DIM) approximation, due to F. O. Ellison and developed by P. J. Kuntz and others. As a *sine qua non,* such trial surfaces must always account for the known overall energetics and diatomic potentials (for AB, BC, *and* CA). One concentrates on achieving the 'proper' qualitative topological shape of the surface and plans on successive variations in key parameters which govern the more quantitative features of the surface, especially the minimum energy path (from the reagents' valley over the barrier to the products' valley).

For a given surface, one then carries out a numerical classical mechanical 'Monte Carlo' computer simulation of the scattering process, sampling the whole range of initial conditions and calculating the fate of a large number of reagent collisions. When one has accumulated a statistically significant number of reactive trajectories, one can obtain directly a measure of the products' vibrational, rotational, and translational energy distributions as well as their angular distribution (which is experimentally available from molecular beam measurements, Chapter 6). Adjustment of the surface and repetition of the calculations can lead to improved, even semiquantitative agreement with experiment, as shown by D. L. Bunker, M. Karplus, J. C. Polanyi, and others.

For a more faithful simulation of state-to-state data, one uses a quasiclassical (rather than purely classical) procedure, effectively 'quantising' both the initial states of the reagent molecules and the final states of the products. There is still a question of the validity of this somewhat ad hoc method, or, from a practical viewpoint, its likelihood of mimicking a quantal scattering calculation. The optimum procedures for such computer simulations as well as validity criteria have recently been thoroughly discussed by D. G. Truhlar and J. T. Muckerman, TRU79.

There are by now many examples in the literature in which experimental state-to-state rate coefficients are 'recovered' by such calculations (which yield thereby empirical, 'best' potential surfaces, 'satisfying' the limited data base). For a few cases in which detailed quantal calculations have been executed on model potentials, the quasiclassical results have indeed exhibited some (but by no means all) of the gross features of the quantal cross sections.

One of the convenient features of the quasiclassical methodology is that one can readily obtain not only angular and product state distributions as a function of initial state and E_{tr}, but also vibrational state-to-state rate coefficients k_{fi} at given T (assuming no vibrational, i.e., only rotational, relaxation). In addition, one obtains directly, by an extension of the R. C. Tolman concept, activation energies for the reaction of selected vibrational states of the reagent molecules, say $E_a^{(v)}$, at a given T, as shown by D. G. Truhlar:

$$E_a^{(v)} \equiv \left[\frac{\text{d}\ln k_v}{\text{d}(1/RT)} \right] = \langle E_{tr} \rangle_{\text{reactive collisions (out of } v)} - \langle E_{tr} \rangle_{\text{all pairs (with } v)} \quad (2.35)$$

The preceding material represents only a glimpse of the subject of chemical reaction dynamics. A great deal more can be found in the literature cited below.

Finally, we can ask, what has all the above to do with molecular beams and lasers? Let us defer this question, postponing further consideration until we have learned a little more (via Chapters 3 and 4) about molecular beam and laser techniques as applied to the study of molecular collisions.

Review literature

GLA41, HAR68, LEV69, ROS69, BUN70, KAR70, POL72, BUN73, CHI74, EYR74, LEV74, NIK74, NIK74a, BER75, EYR75, MIL76, BER77, BRO77, WYA77, BER78, BER79, BER79a, BER79b, FAR79, GLO79, KUN79, KUP79, SCH79, TRU79, EYR80, KNE80, SMI80, ZAR80, JOR81, MOO81, MUC81, TRU81

3 Molecular collisions via crossed beam techniques. Part I: Precollision reagent state preparation

It has become clear that the detailed molecular dynamics of chemical reactions can be elucidated through the use of the crossed molecular beam scattering technique. The present lectures present only a glimpse of the kinds of experiments that provide insights into the microscopic mechanism of chemical reaction. Let us list the unique features of the crossed beam method in its present state of development.

3.1. Unique features of crossed molecular beam techniques

1. *Single collision conditions* are used. This makes it possible to study the outcome of individual collisions of reactant molecules. In a crossed beam experiment, a reagent molecule suffers only one (or none!) collision traversing the beam intersection region. Thus the results are not complicated by effects of sequential (reactive) encounters.

2. *Angular distributions* of the nascent reaction products are directly observed, yielding information about the preferred geometry of the transition state.

3. *Recoil velocity distributions* of the nascent products are measured, i.e., the product flux velocity vector distribution is determined. The products' relative translational energy distribution function indicates the energy partitioning of the reaction.

4. *Velocity selection of reagents* is used. Thus one can study the translational energy dependence of reaction cross sections and of the product flux velocity vector distributions.

5. *Nozzle beam sources yield cold molecules.* By expansion through a nozzle, one can form beams of essentially rotationless molecules, vibrationally cooled as well, and of weakly bound van der Waals 'cluster' molecules as potential reagents for a crossed beam study.

6. *Rotational (and vibrational) state selection* of molecules is possible through the use of inhomogeneous electric and magnetic fields. This allows study of the influence of a reagent's internal energy.

7. *Polarization and orientation* of molecular reagents are achievable. The steric effect can be directly measured through the use of oriented molecule beams.

8. *Rotational state and polarization analysis* of molecular products is possible, yielding information on energy and angular momentum disposal.

9. *Facile interfacing with lasers* is accomplished, allowing for experiments on rotational, vibrational, and electronic excitation of reagent beam molecules and its effect on reactivity. Laser-induced fluorescence of nascent product molecules yields the internal state distribution as a function of scattering angle, providing the most detailed state-to-state cross section data. Polarization of the incident laser beam makes possible measurements of the orientation dependence of the reaction cross section, as above. A crossed beam–laser configuration is ideal for the study of laser-assisted bimolecular reactions.

Before we begin to consider in detail experimental results derived from molecular collision studies, we should try to familiarize ourselves (even, perforce, superficially) with some of the key experimental techniques used to acquire the observational data. It is upon this 'raw' data base that we build the structure of our understanding of chemical dynamics.

The subject can be divided into two parts, the first dealing with the domain of the 'precollision,' i.e., reagent state preparation, the second with 'postcollision,' i.e., product state analysis. In Part I, we shall now consider, successively, the problems of molecular beam formation and concomitant velocity selection (thus determining the relative translational energy), laser excitation of molecules (thus determining reagent internal energy), polarization of laser-excited molecules in beams, and other means of selecting internal energy states of reagent molecule beams. Finally, we shall treat the subject of electrostatic and magnetic focusing and orientation of molecules in beams prior to collision. Postcollision aspects, Part II, are deferred to Chapter 4.

Obviously we can do no more than touch upon each of these subjects in the present cursory presentation. Fortunately there are several excellent reviews on many of these techniques, including those on molecular beams by some of the early practitioners, such as I. Amdur, J. B. Anderson, R. P. Andres, R. B. Bernstein, P. R. Brooks, S. Datz, J. B. Fenn, M. A. Fluendy, E. F. Greene, R. Grice, R. R. Herm, D. R. Herschbach, J. L. Kinsey, G. H. Kwei, K. P. Lawley, Y. T. Lee, J. Los, E. E. Muschlitz, H. Pauly, J. Reuss, J. Ross, E. W. Rothe, C. Schlier, J. P. Toennies, R. N. Zare, and others. Considerable information on techniques can also be found in chapters of various books dealing with modern chemical dynamics (see references at end of chapter). In addition, there is an extensive body of primary literature, to which we shall refer only as necessary. The presentation that follows will be quite subjective (largely because of convenience of illustration), and we make no pretense of a complete, or even systematic, coverage of the subject.

3.2. Apparatus for crossed molecular beam studies

Let us begin by visualizing a 'block diagram' of a hypothetical apparatus intended to carry out a crossed molecular beam experiment to study the reaction

$$A + BC(i) \rightarrow AB(f) + C \tag{3.1}$$

At a minimum we require a vacuum chamber with a couple of separately pumped molecular beam sources, collimated and directed (typically at right angle incidence) so as to intersect one another in a well-defined interaction volume. Each beam should somehow be velocity selected in order to establish a reasonably well-defined (and narrow) relative velocity distribution and thus a known $\overline{E}_{tr}$. Some kind of state selection for the reagent (BC) beam molecules is required as well. All of this is under the heading 'precollision,' of course. We shall now borrow a bit from Chapter 4 and point out the obvious, that we also require a (sensitive!) detector for the scattered products, preferably movable with respect to the incident beams so as to measure the angular distribution of product flux (proportional to the reaction cross section). Moreover, we also desire some means of analysis of the internal state distribution of the scattered product (AB) molecules, or at least the recoil velocity distribution of one of the products, from which the (centre-of-mass) relative translational energy distribution function $P(E'_{tr})$ can be obtained, yielding trivially the 'continuum' internal energy distribution $P(E'_{int})$ of the AB product as a function of its scattering angle.

Figure 3.1 shows a simplified drawing of an actual molecular beam apparatus that incorporates some of these features. Here the A beam is mechanically velocity selected using a Fizeau-type slotted-disk selector (the design and operating principles of which have been discussed by H. U. Hostettler and R. B. Bernstein, and by J. L. Kinsey). The beam of molecules BC issues from a high-pressure nozzle, undergoing a strong isentropic expansion which effectively cools the molecules to a few kelvins. More specifically, the random translational motion of the molecules is converted into directed, forward motion, so that the effective temperature of the beam is reduced to a very low value, with subsequent fast rotational-translational relaxation; vibrational relaxation is slower. For most molecules, we can assume that the highly expanded nozzle beam consists principally of ground (or low-lying) vibrational states, as found by D. H. Levy, L. Wharton, R. E. Smalley, and others. Mainly, however, the Kantrowitz–Grey supersonic seeded nozzle beam technique, made practical by J. B. Fenn, J. B. Anderson, and R. P. Andres, is used because it is a means of producing high-intensity beams with very narrow speed distributions. Also, by using inert, light carrier gases, one can accelerate heavy molecules in a beam. (These points are discussed a little later on.)

Shown in Fig. 3.1, schematically, is the detector, a high-sensitivity electron impact ionization quadrupole mass spectrometer with ion counter, located in a movable ultrahigh vacuum chamber (due to Y. T. Lee, D. R. Herschbach, and others). Such a detector responds linearly to the number density n of the beam molecules traversing the ionizer. A fast chopper wheel is used to create

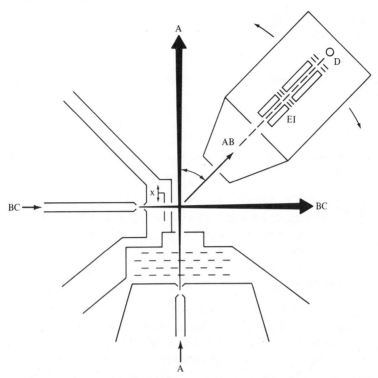

FIG. 3.1. Sketch of a crossed molecular beam machine for the study of reactions such as A + BC → AB + C. The several compartments are separately pumped. The reagent beams, which issue from nozzles, may or may not require velocity selection. The products (e.g., AB) are detected by a sensitive electron impact ionizer–quadrupole mass filter in a rotatable, ultrahigh vacuum chamber. For details, see Y. T. Lee, J. D. McDonald, P. R. LeBreton, and D. R. Herschbach, *Rev. Sci. Instr.* **40**, 1402 (1969); Y. T. Lee and Y. R. Shen, *Physics Today* **33**, 52 (1980); details therein.

pulses of reagent beam molecules, making it possible to measure the time-of-arrival distribution of the beam as well as of the nascent product molecules at the detector, and thus to determine the velocity (v) distribution of product molecules. This allows one to ascertain the product flux I ($\propto nv$) as a function of the scattering angle Θ and to construct a so-called laboratory flux-velocity-angle contour map, say $I(\Theta,v)$, which can be converted from the laboratory to the centre-of-mass system to yield the detailed differential reaction cross sections (to be discussed in Chapter 4).

3.3. Molecular beam sources and beam characteristics

Let us now briefly consider some of the characteristics of molecular beam sources. The early Knudsen oven effusive sources (as described by N. F. Ramsey and P. Kusch) provided beams of low intensity, but well characterised theoretically, e.g., the number density distribution in the beam has been found to be strictly Maxwellian:

$$n(v) \propto v^2 \exp(-v^2/\alpha^2) \; ; \quad \alpha = (2kT/m)^{1/2} \tag{3.2}$$

and the emitted intensity (in molecules per steradian) is found to be exactly as predicted:

$$\frac{dI(\Theta)}{d\omega} = \frac{1}{4\pi} n_0 \bar{v} A \cos\theta \tag{3.2a}$$

so that the total rate of efflux of molecules (in number per unit time) is

$$I = \tfrac{1}{4} n_0 \bar{v} A \tag{3.2b}$$

Here $n_0 = P_0/kT$ is the number density in the oven, P_0 is the pressure in the oven at temperature T, A is the emitting orifice area, θ the angle of emission (with respect to the normal), and $\bar{v} = (8kT/\pi m)^{1/2}$ the average speed of the molecules in the oven.

Note that since the flux of molecules in a mono-velocity beam is $I = nv$, the flux distribution is

$$I(v) = vn(v) \propto v^3 \exp(-v^2/\alpha^2) \tag{3.3}$$

For a nozzle beam (a typical experimental arrangement is shown in Fig. 3.2), it is found empirically, by J. B. Fenn and others, that the number density speed distribution can be well represented by a three-parameter fit:

$$n(v) \propto v^m \exp[-(v - v_s)^2/\alpha_s^2] \tag{3.4}$$

where $m \geq 2$ is an arbitrary exponent, often fixed at 2. Here, v_s is the stream velocity (approximately equal to the most probable velocity in the beam), and α_s is a measure of the width of the speed distribution, governed by the transverse, 'random' component of velocity and serving to define the translational temperature T_s of the beam:

$$\alpha_s = (2kT_s/m)^{1/2} \tag{3.5}$$

The so-called speed ratio of a supersonic beam, S, defined simply as

$$S = v_s/\alpha_s > 1 \tag{3.6}$$

is a measure of the monochromatic quality of the beam. The Mach number M is also used to characterise the beam:

$$M = v_s/(\gamma kT_s/m)^{1/2} = (2/\gamma)^{1/2} S \tag{3.7}$$

where the denominator is the sonic velocity of the molecules at the translational temperature T; $\gamma = C_p/C_v$ is the appropriate specific heat ratio of the beam molecules.

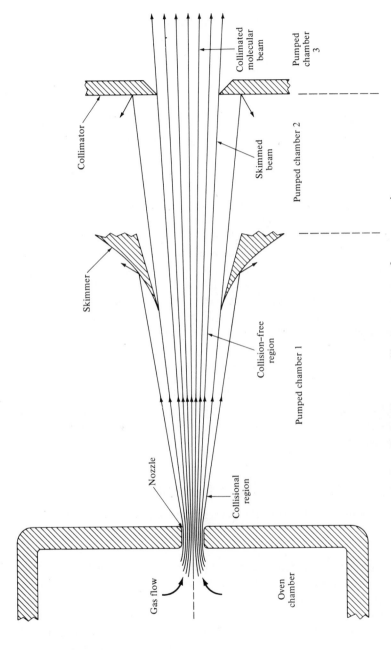

FIG. 3.2. Schematic view of a nozzle-skimmer-collimator source of a supersonic molecular beam. While the pressure in the oven chamber may be of the order of several atmospheres, it must be very much lower in the nozzle-skimmer chamber (no. 1), $< 10^{-3}$ torr, and finally (in no. 3), $\lesssim 10^{-6}$ torr. The distance from nozzle to skimmer is of the order of 10^2 nozzle diameters.

As one increases the degree of the nozzle expansion, governed to a large extent by the product $P_0 d$ (where d is the nozzle diameter), both S and M increase without limit, but the stream velocity of the beam reaches a maximum, or 'terminal,' velocity, say v_t, given by

$$v_t = \alpha[\gamma/(\gamma - 1)]^{1/2} \qquad (3.8)$$

Thus, for a supersonic beam of Ar, for which $\gamma = \frac{5}{3}$, the terminal velocity at 300 K is predicted to be $v_t = 1.58\alpha = 558$ m s^{-1}, in accurate accord with experiment. However, for beams of molecules for which the specific heat and thus γ depend upon temperature, Eqn. 3.8 is not unambiguous. For an expanding jet of polyatomic molecules, the frequent intermolecular collisions ensure that the 'local' rotational temperature follows the translational temperature as it cools, so that as the molecules leave the jet and enter the collisionless region, T_{rot} is only slightly higher than T_s (as shown by L. Wharton, D. H. Levy, R. E. Smalley, J. Jortner, D. R. Herschbach, and others). However, because of the much lower rate of vibrational relaxation and thus of vibrational cooling, the molecules are out of the jet before the vibrational temperature has decreased very much from the original bulk gas temperature, T_0, i.e., T_{vib} 'freezes' out at some value below T_0 but well above T_{rot}. (This makes it difficult to decide on the appropriate γ to use in Eqns. 3.7 and 3.8, but one can make a suitable approximation, as in the 'sudden freeze' model of E. L. Knuth and others.)

If a *mixture* of gases is expanded through the nozzle as a result of the many collisions, the mean molecular weight $\overline{m}$ and mean specific heat ratio $\overline{\gamma}$ are appropriate for Eqns. 3.7 and 3.8. If the mixture consists of a small fraction of a 'heavy' molecule (h), say a possible reagent, 'seeded' in an excess of a light 'carrier' gas (l), the seed molecules will be accelerated to a speed (very nearly) equal to that of the carrier molecules, i.e.,

$$(v_t)_h = (v_t)_l = \left[\frac{2\overline{\gamma}kT}{(\overline{\gamma} - 1)\overline{m}} \right]^{1/2} \qquad (3.9)$$

which, in the limit of the infinitely dilute gaseous solution, yields

$$(v_t)_h = \alpha_l[\gamma_l/(\gamma_l - 1)]^{1/2} \qquad (3.10)$$

Let us see how the seeded beam technique works out. Consider a H_2 carrier gas at 300 K, for which $\alpha = 1.58 \times 10^3$ m s^{-1} and $\gamma = \frac{7}{5}$, so that $v_t = 3 \times 10^3$ m s^{-1}. Thus the average translational energy of the beam molecules in the laboratory framework is $\frac{1}{2}m(v_t)_l^2 = 0.094$ eV. For a seed gas of mass 100 (50-fold that of the carrier), $(v_t)_h \approx (v_t)_l$, so that its translational energy is $\frac{1}{2}m_h(v_t)_l^2 = 50 \times 0.094 \approx 4.7$ eV! By controlling the mole fraction of seed in the carrier, one can readily control the translational energy of the heavy molecule (the reagent) in the seeded supersonic beam. In addition, by increasing the nozzle temperature one can achieve an essentially linear increase in beam translational energy (plus a substantial increase in internal energy, which may or may not be desirable for the given beam experiment).

Through the pioneering efforts of W. R. Gentry, C. F. Giese, and others, intense, *pulsed* supersonic beams are now widely used to produce high but transient (e.g., 10 to 100 μs) fluxes of cold molecules.

In many cases it is found that the cooling in jet formation results in the formation of dimer molecules, e.g., Ar_2 (weakly bound, 'van der Waals molecules') or mixed species ('bimers') such as $Ar \cdots HCl$, $He \cdots I_2$, $Cl_2 \cdots Br_2$, and even large clusters, such as $(CH_3I)_6$ or $(NH_3)_{18}$ (!), from the work of W. Klemperer, D. R. Herschbach, R. B. Bernstein, A. G. Ureña, R. R. Herm, and many others. For very strong expansions one can form microscopic droplets (the onset of condensation) observable by electron and X-ray diffraction. The structure and bonding of these van der Waals molecules is the subject of much current research. Nozzle sources of beams of 'cold molecules' are now widely used, perhaps as much by spectroscopists as by dynamicists!

Returning to the nozzle beam source for molecular scattering experiments, let us recapitulate its principal virtues. These include intensity ($\gtrsim 10^{19}$ molecules s^{-1} sterad^{-1} for cw beams, $\gtrsim 10^{22}$ molecules s^{-1} sterad^{-1} for pulsed beams, whose duty factor is only about 10^{-3}, however), monochromaticity (FWHM of speed distributions typically $\lesssim 10$ per cent), conveniently variable translational energy (range, about 0.05 to 5 eV), and low translational and cold rotational temperatures (typically $\lesssim 10$ K).

Figure 3.3 shows speed distributions for typical supersonic beams of rare gases (or H_2), all of which have essentially reached terminal velocities but with speed ratios S ranging from 5 to 50. For comparison, Fig. 3.4a shows a measured distribution for a moderately expanded beam of SF_6, with speed ratio of 7. Figure 3.4b is a time-of-flight presentation of the same curve; Fig. 3.4c is the observed TOF spectrum. Figure 3.5 is an observed flux-velocity distribution for a very weakly expanded beam of CsF with a low speed ratio.

3.4. Laser excitation of molecules in beams

Next we turn to the subject of laser excitation of molecules in beams. But first we should note the vast body of literature on laser excitation of molecules in the bulk gas phase, making use of photons in the infrared (IR), visible, and ultraviolet (UV). The early work was much restricted by the availability of only line-tunable lasers. At present, even the most advanced designs of continuously tunable lasers provide only limited wavelength ranges.

Despite this problem, a great many valuable experiments on laser excitation of bulk gases have been carried out with the use of the 'workhorse' CO_2 (9.6,

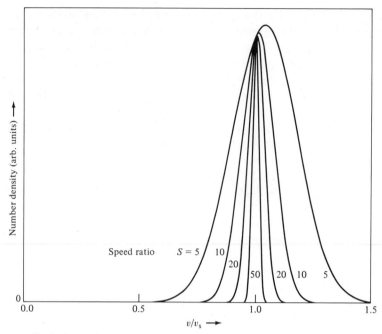

Fig. 3.3. Typical speed distributions, $n(v/v_s)$, for supersonic nozzle beams, with indicated speed ratios $S \equiv v_s/\alpha_s$. Distributions with extremely narrow fractional widths are readily obtainable.

10.4 μm) and HF (2.7 μm), DF (3.8 μm) lasers in the infrared, the Nd YAG laser in the near IR (1.06 μm), the Ar^+ ion (514.5, 488.0 nm) and Kr^+ ion (647.1, 752.5 nm) lasers in the visible, the N_2 (337 nm) laser and the rare gas excimer lasers in the UV, e.g., KrF (249 nm) and ArF (193 nm). With the advent of commercially available tunable dye lasers in the visible and near UV, and of tunable diode and color centre lasers in the infrared, many new and more sophisticated experiments on laser excitation of molecules in bulk gases as well as molecular beams have been carried out.

But we should not overlook the great number of quite plebeian experiments in which all sorts of molecules in the bulk gaseous or liquid phase are irradiated with all sorts of lasers: pulsed and continuous wave (cw); IR, visible, and UV; line-tunable and continuously tunable, etc. Following irradiation, the contents of the cell are analyzed for new products. The goal of most of these experiments is not to probe chemical dynamics but rather to discover new photochemical reactions, to develop a body of 'laser chemistry.'

However, in many cases the laser is used only as a powerful lamp, a collimated source of photons of high fluence; little advantage is taken of the coherence of the photon beam or of the high field strengths present in an intense laser beam. Therefore the results of these experiments are quite often predictable by extrapolation from the well-established notions of organic photochemistry. These usually involve a primary photoexcitation step (whose rate is linear in the intensity of the radiation absorbed) followed by a host of exoergic processes, both uni- and bimolecular, including fluorescence, internal conversion, intersystem crossing, vibrational relaxation, unimolecular decomposition or isomerization, bimolecular reaction, phosphorescence, and excitation transfer.

In some cases, the intensity of the laser beam is sufficient to induce *multi-*

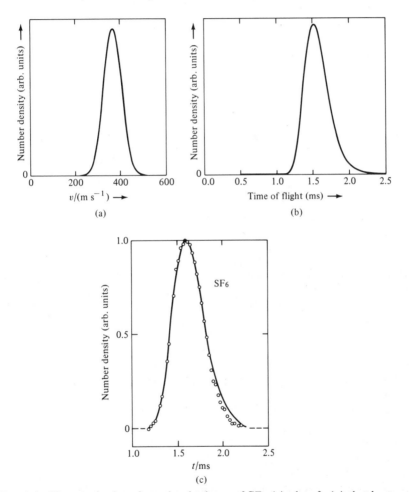

FIG. 3.4. Characterisation of a molecular beam of SF_6: (a) plot of $n(v)$ that best represents the speed distribution; (b) corresponding time-of-flight (TOF) distribution, assuming no instrumental effects; (c) experimental TOF distribution: experimental data points versus solid curves, the convoluted fit to the data of the distribution (b). Adapted from D. R. Coulter, F. R. Grabiner, L. M. Casson, G. W. Flynn, and R. B. Bernstein, *J. Chem. Phys.* **73**, 281 (1980).

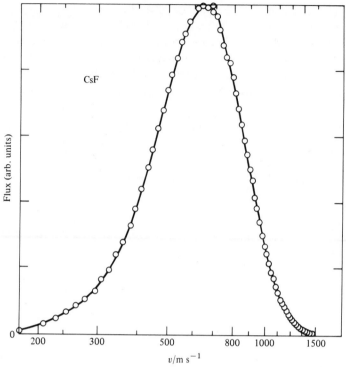

Fɪɢ. 3.5. Flux of CsF from a high-temperature oven source (1270 K) versus log of speed as measured by a slotted-disk velocity analyzer. Data points have been fitted with a slightly modified version of the standard distribution function, Eqn. 3.3, for a speed ratio of 1.15. Adapted from S. Stolte, A. E. Proctor, and R. B. Bernstein, *J. Chem. Phys.* **65**, 4990 (1976).

photon excitation of the absorbing molecules (to be discussed in Chapter 10). In particular, the pulsed CO_2 laser can provide such extremely intense IR photon beams (in the 9.6 to 10.4 μm region) that nearly every molecule (with an absorption band overlaying one of the laser line) in the laser beam can be dissociated during a single laser pulse. In this and other multiphoton processes, the dependence upon light intensity is usually much stronger than linear, the exact behaviour being governed by the molecular energy level up-pumping mechanism. The relative importance of instantaneous peak power density (photons m^{-2} s^{-1}) versus total fluence (photons m^{-2} per pulse) upon the photodissociation rate is a subject of current interest. The field of infrared laser photochemistry is a rapidly growing one (see Chapter 10).

    Another interesting aspect of laser excitation is the utilization of the high photon fluxes to effect weak, or 'forbidden,' transitions from the ground to excited vibronic states. These experiments may be carried out intracavity (i.e., with the weakly absorbing sample cell located within the laser resonant cavity, where photon densities may be several orders of magnitude greater than in the transmitted laser beam). Intracavity cw dye laser excitation of gases has made it possible to study both the spectroscopy and the reactions of highly vibrationally excited states of polyatomic molecules (as shown by M. J. Berry and coworkers). The high photon fluxes available intracavity and the use of a sensitive photoacoustic detector allow observation of 'highly forbidden' vibronic transitions. Also, direct one-photon absorption can produce significant concentrations of overtone-excited molecules (whose reactivity may be enhanced with respect to the ground state molecules).

Well-resolved one-photon absorption spectra have been recorded for HCl, corresponding to transitions from $v = 0$ to 5, 6, and 7 (near 747, 634, and 554 nm, respectively), high overtone absorptions for benzene, C_6H_6 ($5\nu_{CH}$, $6\nu_{CH}$, and $7\nu_{CH}$), and for methyl isocyanide, CH_3NC ($5\nu_{CH}$, $6\nu_{CH}$, and $7\nu_{CH}$). Experiments on the rate of the unimolecular isomerization reaction ($CH_3NC \rightarrow CH_3CN$) showed successively larger enhancements in rate for $5\nu_{CH}$, $6\nu_{CH}$, and $7\nu_{CH}$ excited reagent molecules (in approximate accord with unimolecular reaction rate theory). K. V. Reddy and M. J. Berry have excited allyl isocyanide ($CH_2{=}CHCH_2NC$) selectively, e.g., $6\nu_{CH}$ (methylenic), $6\nu_{CH}$ (olefinic CH), and $6\nu_{CH}$ (olefinic CH_2), and have observed the resulting unimolecular isomerization rates, showing some evidence for a bond-selective excitation effect.

Intracavity excitation has yielded specific vibrational states of singlet oxygen molecules:

$$O_2(X^3\Sigma_g^-) \xrightarrow[\lambda762\ nm]{h\nu} O_2^*(b^1\Sigma_g^+) \qquad (\Delta E_0 = 1.64\ \text{eV}) \qquad (3.11)$$

Tunable dye laser excitation of oxygen gas (to the $v = 0$ level of the $b^1\Sigma_g^+$ state), in the presence of excess olefins, leads to new addition reactions. Unfortunately, the results of such bulk experiments do not preclude collisional steps, such as internal conversion or intersystem crossing to convert the $O_2^*(^1\Sigma_g^+)$ to $O_2^*(a^1\Delta_g)$, ($\Delta E_0 = -0.66$ eV) or highly vibrationally excited $O_2(^3\Sigma_g^-)$, either of which may react with the olefin. Once the $^1\Sigma$ oxygen is quenched to the $^1\Delta$ manifold, further chemistry would be that of $^1\Delta_g\ O_2$. Thus laser excitation studies in the bulk gas phase are often equivocal because of concurrent collisional relaxation processes. Beam experiments are inherently less ambiguous and therefore advantageous.

The earliest studies of electronic excitation in beams deal with the I_2 molecule. Argon ion laser excitation yields the B state:

$$I_2\left(X^1\Sigma_g^+, v'' = 0 \right) \xrightarrow[\lambda514.5\ nm]{h\nu} I_2(B^3\Pi_{0+u}, v' = 43) \qquad (3.12)$$

The fluorescence from a laser-excited molecular beam of I_2 can be observed.

The rate of the competitive direct predissociation process

$$I_2^*(^1\Pi_u) \rightarrow I(^2P) + I(^2P_{32}), I(^2P_{1/2}) \tag{3.13}$$

is also known, as are the electronic transition moments of the B $\rightarrow$ X system of I_2 (via gain measurements of an optically pumped I_2 laser). Spectroscopic data and Franck–Condon factors are also available. The tunable-laser-induced fluorescence technique discussed in Chapter 4 can be used to determine vibrational and rotational state distributions of I_2 (in seeded, supersonic molecular beams). This allows one to estimate optimal beam-formation conditions for achieving the maximum number density of I_2^* in a molecular beam, for use as a reagent in crossed beam collision studies. The factor limiting the degree of laser excitation is not the available laser power (saturation is readily achieved) but rather the population of the given X-state rovibrational level to be excited to the B state.

The Ar^+ ion laser line (multimode) excites essentially only the P(13) and R(15), 43'-0'' transitions, so that the initial population of the $v'' = 0$, $J'' = 13, 15$ states determines the final I_2^* population in the beam. Assuming a Boltzmann internal state distribution, the rotational temperature which maximizes the fraction of $J'' = 13$ is about 10 K, at which the fraction $f(J'' = 13) \approx 0.055$; $f(J'' = 15) \approx 0.048$. The 21 hyperfine components of each of the lines, slightly Doppler broadened, are overlaid by some 16 or so laser modes. Experiments on the dependence of the I_2^* fluorescence on the laser power by M. A. McMahan, M. M. Oprysko, F. J. Aoiz, and R. B. Bernstein indicate the onset of saturation at a few watts of incident (multimode) power. This has been confirmed by direct measurement of the laser-induced loss of I_2 in the beam via Eqn. 3.13. Thus I_2 beam-excitation experiments are limited more by the partition function, the I_2^* predissociation problem, and the short ($\approx 3 \mu s$) radiative lifetime than by subtle, 'fine-tuning' considerations. Under the best conditions thus far, only about 1 per cent of the I_2 molecules in a beam have been excited to the B state.

3.5. Vibrationally excited molecule beams

Next let us consider the formation of *vibrationally* excited molecules in beams. Excited hydrogen halide (HX) molecules have been produced by resonant excitation (by hydrogen halide–laser irradiation of collision-free beams of HX molecules). P. R. Brooks and co-workers showed this in connection with their study of the reaction

$$HCl(v = 1) + K \rightarrow KCl + H \tag{3.14}$$

More recent experiments have involved J-state selection of the HCl as well, using a grating-tuned HCl chemical laser for irradiation of the HCl beam.

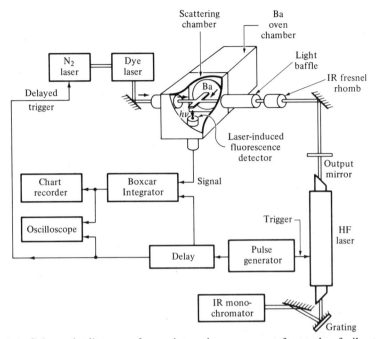

FIG. 3.6. Schematic diagram of experimental arrangement for study of vibrational enchancement of reactivity. HF (contained in a scattering chamber) is resonantly excited by a pulsed HF laser and can then react with an atomic beam of Ba or Sr. A tunable dye laser probes the nascent product state distribution by inducing fluorescence, which is monitored by a photomultiplier. Adapted from Z. Karny and R. N. Zare, *J. Chem. Phys.* **68**, 3360 (1978); Z. Karny, R. C. Estler, and R. N. Zare, *J. Chem. Phys.* **69**, 5199 (1978); details therein.

R. N. Zare and co-workers have produced $HF(v = 1)$ in a scattering chamber (here not a molecular beam) with a pulsed HF (TEA) laser (with a polarizer for the laser beam) and studied the reactions

$$HF^\dagger + Ba, Sr \rightarrow BaF, SrF + H \qquad (3.15)$$

using the apparatus sketched in Fig. 3.6.

Finally, let us consider the use of the chemical laser for resonant excitation of beam molecules. Figure 3.7 shows a sketch of a cryosorption-pumped, fast-flow, cw chemical laser for HCl and HF excitation. Its output is presented in Fig. 3.8. It provides a 200-mW output at the $P_1(5)$ line of HCl and 300 mW at the $P_1(4)$ line of HF, suitable for beam excitation.

The explanation of the operation of such a chemical laser is as follows. For the sake of definiteness, let us consider the particular 'lasing reaction'

$$Cl + HBr \rightarrow HCl^\dagger + Br \qquad (3.16)$$

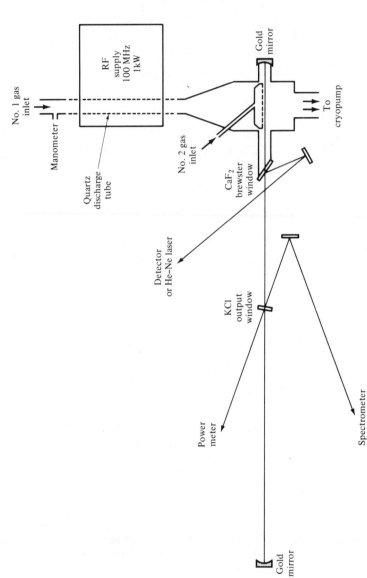

FIG. 3.7. Sketch of cryosorption-pumped cw hydrogen halide chemical laser. The RF discharge occurring in a stream of a suitable precursor gas, no. 1, produces halogen atoms, which react rapidly as they mix with a stream of the appropriate co-reagent gas, no. 2, within the cavity, resulting in multiline lasing. Adapted from K. R. Newton and R. B. Bernstein, *Appl. Phys. Lett.* **33**, 47 (1978); details therein.

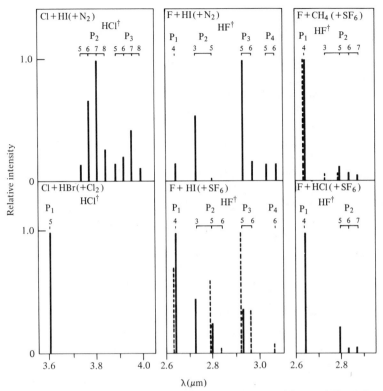

FIG. 3.8. Multiline IR output from cryopumped cw chemical laser of Fig. 3.7; reference therein.

known to produce an 'excess population' of HCl in the state $v' = 1$ (levels with $v' \geq 2$ are energetically inaccessible). In the apparatus of Fig. 3.7, the Cl atoms (formed in the discharge by the dissociation of Cl_2) in a N_2 carrier gas flow down a tube and react rapidly with HBr molecules issuing from the many small orifices (spaced along an injection length of 0.1 m transverse to the flow direction). The reaction produces many $HCl^{\dagger}$ molecules, with a substantial fraction in the state $v' = 1$, $j' = 4$, all of which are being rapidly pumped away. Some of these will fluoresce while traversing the cavity, and the emitted IR photon from one of them can stimulate the coherent emission of another photon of the same frequency from a nearby HCl($v' = 1$, $j' = 4$) molecule to yield two 'coherent' photons. It will always be the case that the direction of at least one of those first-emitted photons happens to lie more or less along the optical axis (within the cavity, or 'optical resonator,' defined by the two end mirrors), so that there will be an exponential gain in the intensity of the coherent radiation as a function of distance along the active length of the cavity, offset by a loss term resulting from the absorption of these photons by some HCl molecules in the ground vibrational state (specifically, $j' = 5$) which have not yet been swept out of the cavity. The active length of the cavity is limited by the injection length, here only 10 per cent of the optical cavity length. The higher the flow rates and the faster the population-inverting reaction the higher the net optical gain and the smaller the loss. If one of the mirrors is replaced by a grating, one can select a particular transition on which to lase. In the arrangement shown, the laser is in general a multiline source, but for the example chosen it lases on only a single line, the $P_1(5)$ line, corresponding to the transition in HCl from $v = 1$, $j = 4$ to $v = 0$, $j = 5$ (at $\lambda = 3.61$ μm).

It is of interest to estimate the fractional excitation of a molecular beam, say a thermal (300 K) HCl beam, by the 200-mW laser output. For such a Boltzmann beam, 12.3 per cent of the molecules are in the state ($v = 0$, $j = 5$) which can absorb the $P_1(5)$ line radiation. The fractional excitation f of these molecules can be estimated using a 'two-level' treatment.

Let n_1 and n_2 be the number densities of the lower and upper states, respectively, and $n \equiv n_1 + n_2$ the total; then the fractional exitation $f_2 = n_2/n$ is given by

$$f_2 = \frac{g_2}{g_1 + g_2} \left\{ 1 - \exp\left[-\rho B_{21}\left(\frac{g_1 + g_2}{g_1} \right) t \right] \right\} \qquad (3.17)$$

where g_1 and g_2 are the rotational degeneracies of the ground and excited states and

$$\rho(\nu) = P/(Ac^2\Delta\nu) \qquad (3.18)$$

is the radiation density at the transition frequency ν. Here P is the laser power, A the laser beam area (assumed to match that of the molecular beam), $\Delta\nu$ the laser linewidth, and c the speed of light. The B_{21} term is the Einstein stimulated emission coefficient, calculated from A_{21}, the spontaneous emission coefficient, via the relation

$$B_{21} = A_{21}/8\pi h\tilde{\nu}^3 \qquad (3.19)$$

where $\tilde{\nu} = \nu/c$ is the wavenumber and h is Planck's constant. The exposure time t can be estimated from the mean molecular speed $\bar{v} = 1.128\alpha = 4.17 \times 10^2$ m s^{-1} and the length of the irradiation zone, say 10^{-2} m, to be about 2.4 μs. The line width $\Delta\nu$ is governed by the Doppler width of the target molecules:

$$\frac{\Delta\tilde{\nu}}{\tilde{\nu}} = \frac{\Delta\nu}{\nu} \approx \langle v_{\text{transverse}} \rangle / c \approx [2(\ln 2)^{1/2}]\alpha/c \qquad (3.20)$$

where the numerator is the FWHM of the transverse component of the molecular speed. For HCl at 300 K, using $\tilde{\nu} = 2775.8$ cm^{-1}, Eqn. 3.20 gives $\Delta\tilde{\nu} =$

5.7×10^{-3} cm^{-1}. For an area A of 10^{-2} cm^2, Eqn. 3.18 yields $\rho = 3.9 \times 10^{-18}$ J s cm^{-3}. From the known values of A_{21} for the transition, namely 19.2 s^{-1}, Eqn. 3.19 yields $B_{21} = 5.4 \times 10^{22}$ cm^3 J^{-1}s^{-2}. Thus the fractional excitation (Eqn. 3.17) can be estimated to be $f_2 = \%_0[1 - \exp(-3.9 \times 10^{-18} \times 5.4 \times 10^{22} \times {}^{20}\!/_{11} \times 2.4 \times 10^{-6}] = \%_0 \times 0.60 = 0.27$. (The number 0.60 implies 60 per cent 'saturation' of the transition. Thus there is little to be gained by higher laser powers.) Thus 27 per cent of the 12.3 per cent of the beam molecules originally in the $v = 0, j = 5$ state are excited. Overall, therefore, 3.3 per cent of the total beam is excited to the state $v = 1, j = 4$ under the conditions described.

Once again, as in the case of $I_2(X \rightarrow B)$ laser excitation, the main limitation is that due to the partition function, not the limited photon flux density of the laser.

Before leaving the subject of laser excitation, it seems worthwhile to note the magnitude of the photon energetics and fluxes for typical pulsed and cw lasers in the IR, visible, and UV regions. We recall that the energy 'content' of a photon, say ϵ, is

$$\epsilon = h\nu = hc/\lambda = hc\tilde{\nu} \qquad (3.21)$$

where $\tilde{\nu}/$cm^{-1} is the wavenumber. In joules per photon,

$$\epsilon/\text{J} = 1.9865 \times 10^{-23} \tilde{\nu}/\text{cm}^{-1} = 1.9865 \times 10^{-16}/(\lambda/\text{nm}) \qquad (3.22)$$

Thus the number of photons per laser pulse, say N, corresponding to j joules per pulse is

$$N = 5.034 \times 10^{15}(j/\text{J})(\lambda/\text{nm}) \qquad (3.23)$$

For cw operation, the intensity in number of photons s^{-1}, say I, corresponding to a power of P watts is

$$I/\text{s}^{-1} = 5.034 \times 10^{15}(P/\text{W})(\lambda/\text{nm}) \qquad (3.24)$$

Let us now go through a few numbers for three wavelengths.

(A) $\lambda = 10.4$ μm (1.04×10^4 nm), one of the strong IR lines from a CO_2 laser. Here $\epsilon = 1.91 \times 10^{-20}$ J (0.119 eV). For a cw laser delivering a power of 100 watts, $I = 5.24 \times 10^{21}$ photons s^{-1}. For a pulsed laser providing 2 J pulse^{-1}, $N = 1.05 \times 10^{20}$ photons pulse^{-1}; for a pulse duration of 0.1 μs, this corresponds to an average wattage of 20 MW and an average intensity of some 10^{27} photons s^{-1}! To obtain an Einstein (N_{Avog} of photons), one needs only 5.7×10^3 pulses.

(B) $\lambda = 560$ nm, a typical visible wavelength obtainable from a tunable dye laser (e.g., a UV laser-pumped rhodamine dye system). Here $\epsilon = 3.55 \times 10^{-19}$ J (2.22 eV). For 10-mJ pulses, $N = 2.82 \times 10^{16}$ photons pulse^{-1}; for a pulse duration of 10 ns, $\overline{P} = 1$ MW and $\overline{I} = 2.82 \times 10^{24}$ photons s^{-1}. For a 10-ps pulse, $\overline{P} = 1$ GW (and $\overline{I} = 2.82 \times 10^{24}$ photons s^{-1}). For a cw laser delivering 500 mW, $I = 1.41 \times 10^{18}$ photons s^{-1}.

(C) $\lambda = 193$ nm, the prominent line from an ArF UV excimer laser. Here, $\epsilon = 1.03 \times 10^{-18}$ J (6.43 eV). For a laser yielding 50 mJ pulse^{-1}, $N = 4.86 \times 10^{16}$ photons pulse^{-1} (requiring 1.24×10^{7} pulses for an Einstein).

Enough of these details. Suffice it to say that improvements in laser technology will soon make these numbers obsolete. Higher peak power pulsed lasers of higher repetition rates (higher duty factors) as well as higher power cw lasers are imminent, and so laser excitation experiments will be much easier, and therefore prevalent, in the future.

3.6. State selection in molecular beams

Let us return now to the main subject of this chapter, namely, precollision beam preparation. We have just seen how selective *excitation* of molecules in beams (by laser irradiation) is carried out. However, for quantitative application of this technique to scattering experiments, one requires a knowledge of the fractional excitation of the beam molecules. It is also implicit that we are dealing with a collision process especially sensitive to the newly won excited state. An alternative approach is to *select* from among the beam molecules (i.e., choose from a thermal beam with a Boltzmann internal state distribution) certain states of interest and perform the scattering experiment with the given selected state (or a known 'mixture' of such states). This is especially desirable for experiments on rotational effects for ground vibrational state molecules. (Such experiments deal with the influence of the rotational angular momentum and its polarization, as well as rotational energy per se, upon the cross section for a variety of collisional processes.)

The method of state selection to be briefly outlined below makes use of the fact that certain classes of molecules can be deflected in flight by inhomogeneous electrostatic or magnetic fields (as shown by W. Paul, H. G. Bennewitz, and C. Schlier). For example, polar molecules (which exhibit a Stark effect) can be rotationally state-selected by the use of an electric quadrupole field. For diatomics, both the rotational quantum number j and the orientation quantum number m_j can be selected, by virtue of their second-order Stark effect (e.g., by L. Wharton, W. Klemperer, and others). Thus it is possible to work with beam molecules of known rotational energy and polarization with respect to an alignment field. (In the classical limit, the rotating diatomic is a disk whose normal is processing at a given angle with respect to the field.) Symmetric-top and asymmetric-top molecules, because of their first-order Stark effect, are easily deflected and can be partially state-selected by electric hexapole fields (e.g., by K. H. Kramer and R. B. Bernstein and by P. R. Brooks and co-workers). Nonpolar molecules, especially H_2, have been state-selected and beams of polarized molecules formed by the use of inhomogeneous magnetic fields (e.g., by J. Reuss and co-workers). Similarly, beams of odd-electron species, including atoms, radicals, and the NO molecule, have been state-selected through the use of magnetic hexapole fields (e.g., by S. Stolte and others).

In addition to state selection, it is possible to *focus* molecular beams with suitable multipole fields. The field assembly has an action akin to that of a thick lens (as elaborated by L. Wharton), accepting a cone of molecules issuing from a small source and focusing the molecules in given selected states down to a small image of the source at the focal point of the lens. There one can perform a variety of scattering experiments on the state-selected beam, some of which will be described later. At this point, it is worth mentioning that the state-selected molecules may, under certain conditions, be oriented (not merely aligned, i.e., polarized).

3.7. Focusing and state selection via inhomogeneous electric fields

Let us briefly consider the principles underlying the state selection and focusing technique. For simplicity, we shall limit our attention to electrostatic (not magnetic) focusing. We recall that the Stark effect for molecules induces field-dependent splittings of degenerate rotational levels. Without going into details, we write down the perturbation series for the energy of a polar molecule in a given rotational state, say E_{JKM}, as a function of the field strength $\mathscr{E}$:

$$E_{JKM} = E^{(0)} + E^{(1)} + E^{(2)} + \cdots = E_{JKM}^0 + \lambda_1\mathscr{E} + \lambda_2\mathscr{E}^2 + \cdots \quad (3.25)$$

where $E_{JKM}^{(0)}$ is the energy of the unperturbed rotor in the limit of zero field strength and λ_1 and λ_2 are the coefficients of the first- and second-order Stark terms, each dependent upon molecular parameters and rotational quantum numbers.

To make contact with the classical picture, we imagine the molecule's electric dipole moment μ located within a field $\mathscr{E}$, as shown below:

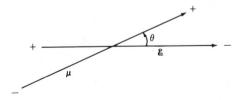

Letting W be its energy, then classically we have

$$W = W^{(0)} - \mu \cdot \mathscr{E} = W^{(0)} - \mu \cos\theta \; \mathscr{E} \quad (3.26)$$

where $W^{(0)}$ is the field-free value. Influenced by Eqn. 3.26, we define the 'effective' dipole moment of the molecule in the field to be

$$\mu_{\text{eff}} = \mu\langle\cos\theta\rangle = -\frac{\partial W}{\partial \mathcal{E}} \qquad (3.27)$$

where $\langle\cos\theta\rangle$ is the quantum mechanical expectation value of the cosine of the angle of orientation of the dipole moment with respect to the field. From Eqns. 3.25 and 3.26,

$$\mu_{\text{eff}} = -\lambda_1 - 2\lambda_2\mathcal{E} + \cdots \qquad (3.28)$$

The force on the molecule at some position r in the field is given classically by

$$F_r = -\frac{\partial W}{\partial r} = -\frac{\partial W}{\partial \mathcal{E}}\frac{\partial \mathcal{E}}{\partial r} = \mu_{\text{eff}}\frac{\partial \mathcal{E}}{\partial r} \qquad (3.29)$$

(where we have used Eqn. 3.27) so that, from Eqn. 3.28,

$$F_r = -\lambda_1\frac{\partial \mathcal{E}}{\partial r} - \lambda_2\mathcal{E}\frac{\partial \mathcal{E}}{\partial r} + \cdots \qquad (3.30)$$

Let us consider a symmetric-top molecule, for which the first-order term in the Stark series (Eqn. 3.25) suffices (at not-too-high field strengths). Then we have $E_{JKM} \approx E^0_{JKM} + E^{(1)}$, with

$$E^{(1)} = -\mu_{\text{eff}}\mathcal{E} = -\mu\langle\cos\theta\rangle\mathcal{E} = \lambda_1\mathcal{E} \qquad (3.31)$$

From quantum mechanics, the first-order Stark energy is known to be

$$E^{(1)} = -\mu\frac{KM}{J(J+1)}\mathcal{E} \qquad (3.32)$$

and so by comparison with Eqn. 3.31, we make the identification

$$\langle\cos\theta\rangle = KM/(J^2 + J) \qquad (3.33)$$

(Note that this measure of the alignment of the dipole is independent of the magnitude of the field, in this first-order limit.)

Thus Eqn. 3.29 can be expressed in the form

$$F_r \approx -\lambda_1\frac{\partial \mathcal{E}}{\partial r} = \frac{KM}{J^2 + J}\mu\frac{\partial \mathcal{E}}{\partial r} \qquad (3.34)$$

We shall make use of Eqn. 3.34 shortly, but first let us consider the case of the polar diatomic (or linear polyatomic) molecule, for which $E^{(1)} = 0$ (i.e., there is no first-order Stark effect, $\lambda_1 = 0$; the leading term in the rotor energy is quadratic in the field). Thus Eqn. 3.25 becomes

$$E_{JM} = E^0_{JM} + \lambda_2 \mathcal{E}^2 + \cdots \tag{3.35}$$

where $E^0_{JM} = BJ(J + 1)$, for a rigid rotor ($B = \hbar^2/2I$), and

$$\lambda_2 = f(J,M)\frac{\mu^2}{B} \tag{3.36}$$

with $f(J,M)$ a known function of the quantum numbers.* Retaining the definition of μ_{eff} of Eqn. 3.27, we have

$$\mu_{\text{eff}} \equiv \mu\langle\cos\theta\rangle \approx -2\lambda_2\mathcal{E} = -2\frac{\mu^2}{B}f(J,M)\mathcal{E} \tag{3.37}$$

so that

$$\langle\cos\theta\rangle = -2f(J,M)\frac{\mu\mathcal{E}}{B} + \cdots \tag{3.38}$$

(Note that, in contrast to the symmetric-top case, Eqn. 3.33, here the 'alignment' for a given quantum state depends linearly on the field strength.) Equation 3.30 becomes (using Eqn. 3.36)

$$F_r \approx -2\lambda_2\mathcal{E}\frac{\partial\mathcal{E}}{\partial r} = -2f(J,M)\frac{\mu^2}{B}\mathcal{E}\frac{\partial\mathcal{E}}{\partial r} \tag{3.39}$$

to be compared with Eqn. 3.34 for the symmetric-top case.

Next let us place our molecules in a multipole electric field. Shown in Fig. 3.9 are idealized hyperbolic four-pole and six-pole fields for which the field strengths are of especially simple form. Table 3.1 compares their relevant properties. Thus for a polar diatomic molecule in a quadrupole electric field (Case

TABLE 3.1

Property	Four-pole field	Six-pole field
$V(r,\phi)$	$V_0\left(\dfrac{r}{r_0}\right)^2 \cos 2\phi$	$V_0\left(\dfrac{r}{r_0}\right)^3 \cos 3\phi$
$\mathcal{E}_r$	$\left(\dfrac{2V_0}{r_0^2}\right)r$	$\left(\dfrac{3V_0}{r_0^3}\right)r^2$
$\dfrac{\partial\mathcal{E}_r}{\partial r}$	$\dfrac{2V_0}{r_0^2}$	$\left(\dfrac{6V_0}{r_0^3}\right)r$

*$f(J,M) = (J^2 + J - 3M^2)/[2(J^2 + J)(2J - 1)(2J + 3)]$ except for $J = M = 0$, for which $f(0,0) = -\frac{1}{6}$.

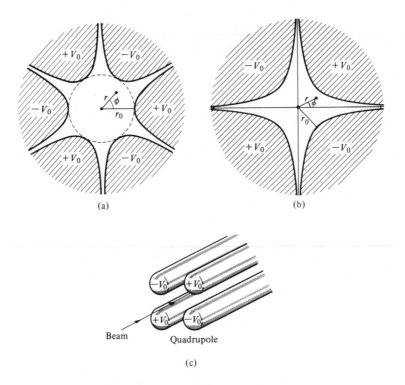

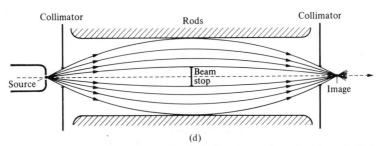

FIG. 3.9. (a) Ideal electric hexapole; (b) ideal electric quadrupole; (c) typical electric quadrupole rod assembly; (d) stylized molecular trajectories, showing focusing effect of the field (rod) assembly; note analogy with 'thick lens.'

A), the radial force (Eqn. 3.39) is given by

$$F_r = -2f(J,M) \frac{\mu^2}{B} \frac{4V_0^2}{r_0^4} r = -k_A r \qquad (3.40a)$$

where

$$k_A = 8 \frac{V_0^2}{r_0^4} \frac{\mu^2}{B} f(J,M) \qquad (3.40b)$$

is a force constant for radial motion.

For a symmetric-top molecule in a hexapole (Case B), the radial force (Eqn. 3.34) is

$$F_r = \frac{KM}{J(J+1)} \mu \frac{6V_0}{r_0^3} r = -k_B r \qquad (3.41a)$$

where

$$k_B = -6 \frac{V_0}{r_0^3} \mu \frac{KM}{J(J+1)} \qquad (3.41b)$$

is, similarly, the radial force constant.

Now consider a beam of molecules moving down the z axis of the field (of length l) with a speed v, injected at various slight angles from the axis. The field has no effect on the longitudinal motion but acts upon the molecules' radial component of velocity. Using Eqns. 3.40 or 3.41, Newton's law gives us

$$F_r = -kr = m\ddot{r} \qquad (3.42)$$

a differential equation with a sinusoidal (if $k > 0$) or exponential (if $k < 0$) solution. For $k > 0$, with the initial condition $r = 0$ at $z = 0$, Eqn. 3.42 yields, for $z = l$,

$$r = \sin[(k/m)^{1/2}l/v] \qquad (3.43)$$

so that the molecule will follow a sinusoidal path and be returned to the axis when $(k/m)^{1/2}l/v = n\pi$ ($n = 1, 2, 3, \ldots$).

For a single-loop trajectory, $n = 1$ and $k = \pi^2 mv^2/l^2$, so that for Case A, using Eqn. 3.40b, the condition for focusing can be written

$$\frac{8V_0^2\mu^2 f(J,M)}{r_0^4 B} = \frac{\pi^2 mv^2}{l^2} \qquad (3.44a)$$

or

$$V_0^2 = \frac{\pi^2}{8} \frac{r_0^4}{l^2} \frac{mv^2}{(\mu^2/B)} \frac{1}{f(J,M)} \qquad \text{(Case A)} \qquad (3.44b)$$

(Note that only states for which $f(JM) > 0$ will be focused; cf. Eqn. 3.42.)

For Case B, using Eqn. 3.41b, similarly

$$-6 \frac{V_0}{r_0^3} \mu \frac{KM}{J(J+1)} = \frac{\pi^2 m v^2}{l^2} \qquad (3.45a)$$

or

$$V_0 = \frac{\pi^2}{6} \frac{r_0^3}{l^2} \frac{m v^2}{\mu} \left[\frac{-KM}{J(J+1)} \right]^{-1} \qquad \text{(Case B)} \qquad (3.45b)$$

(Once again, note that the only states that can be focused are those for which $\langle \cos\theta \rangle < 0$.)

Thus, given the molecular parameters m, B, μ, and the speed of the beam molecules v, and knowing the field dimensions (l and r_0), Eqns. 3.44b and 3.45b make it possible to calculate the 'rod voltage' V_0 at which selected rotational states will be focused at the image point.

Thus, by 'scanning' the rod voltage from zero, typically, to about 15 KV, different rotational states of the beam molecules are successively brought into focus and may be used as a source of state-selected molecules, as shown by K. H. Kramer and R. B. Bernstein.

Experimentally, one usually eliminates the direct, unfocused beam ($n = 0$) and the even-loop trajectories ($n = 2, 4, \ldots$) by interposing a beam stop at $z = l/2$, as shown in Fig. 3.9d. The higher odd ($n = 3, 5, \ldots$) trajectories tend to diverge rapidly past the first image collimator and contribute negligibly to the final, near-axial, collimated beam.

3.8. Experimental aspects of state selection

For Case A, i.e., diatomics focused by a quadrupole, it is possible to isolate successively several of the lower rotational states (restricted, of course, to those for which $\mu_{eff} < 0$, i.e., $f(J,M) > 0$), as first shown by H. G. Bennewitz, C. Schlier, K. H. Kramer, and J. P. Toennies and later by L. Wharton and by R. B. Bernstein and co-workers T. G. Waech, R. W. Fenstermaker, E. E. Bromberg, and A. E. Proctor. Figure 3.10 shows a typical 'focusing curve' for a CsF beam, showing the states $J,M = 1,0; 2,0$; higher states are here not well resolved. Figure 3.11 shows an apparatus for scattering of selected rotational states of alkali halide molecules.

Because of the large partition function at the high oven temperature, only a small fraction of the beam molecules are in these low rotational states. The fraction in state J,M is

$$F_{J,M} = \frac{\exp(-E_{J,M}/kT)}{Q_{rot}} \approx \frac{B}{kT} \ll 1 \qquad (3.46)$$

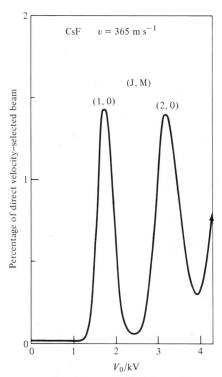

FIG. 3.10. Typical focusing curve for velocity-selected beam of CsF using electric quadrupole field, showing separation of J,M states 1,0 and 2,0. Adapted from E. E. Bromberg, A. E. Proctor, and R. B. Bernstein, *J. Chem. Phys.* **63**, 3287 (1975); details therein.

the approximation valid for low J and high T (when $E_{rot} \ll kT$). Despite this small fraction, the large aperture of the 'thick lens' system of the quadrupole allows for a considerable enhancement in flux of the selected state over that of the same state in the unfocused beam, i.e., that which would be received at the final collimator if the beam stop were removed.

Finally, for Case B, i.e., symmetric tops focused by a hexapole, Eqn. 3.45b shows that the state selection is according to the combination of quantum numbers $KM/(J^2 + J)$, equivalent to $\langle \cos\theta \rangle$, i.e.,

$$V_0 = V_{th}/(-\langle \cos\theta \rangle) \qquad (-1 \le \langle \cos\theta \rangle < 0) \qquad (3.47)$$

where V_{th} is found from Eqn. 3.45b. A given V_0 supplies beam molecules with a given $\langle \cos\theta \rangle$; as V_0 is reduced in magnitude from a high voltage for which the precession angle is approximately $90°$ (i.e., $\langle \cos\theta \rangle \approx 0$) towards a lower

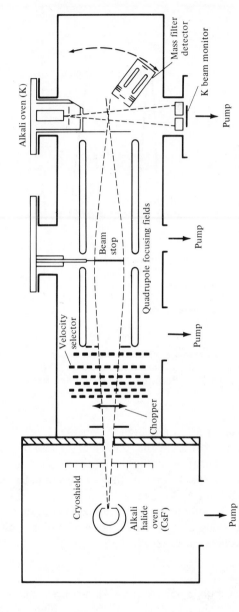

FIG. 3.11. Schematic of crossed molecular beam apparatus for study of reactions of state-selected alkali halide molecules with alkalis. Adapted from S. Stolte, A. E. Proctor, W. M. Pope, and R. B. Bernstein, *J. Chem. Phys.* **66**, 3468 (1977); details therein.

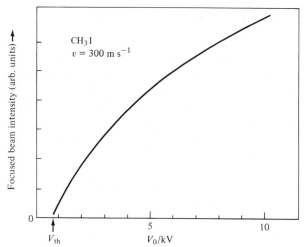

FIG. 3.12. Focusing curves for velocity-selected beam of CH_3I using electric hexapole field. Adapted from K. H. Kramer and R. B. Bernstein, *J. Chem. Phys.* **42**, 767 (1965); details therein.

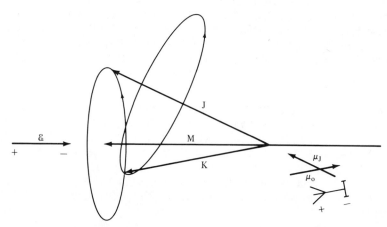

FIG. 3.13. Precession of symmetric-top molecule (e.g., CH_3I) in specified J,K,M states in an electric field. See R. J. Beuhler and R. B. Bernstein, *J. Chem. Phys.* **51**, 5305 (1969) for details; oppposite sign convention for μ_o used, however.

value, the precession angle becomes smaller until eventually $\langle \cos\theta \rangle = -1$ when V_0 is reduced to a 'threshold voltage' V_{th}, below which no molecules are focused.

Figure 3.12 shows a typical 'focusing curve' for a symmetric-top molecule, i.e., a plot of the intensity of the focused beam as a function of the rod voltage. For a given V_0 there is a known value of $\langle \cos\theta \rangle$. These molecules passing out through the hexapole can then be oriented in the laboratory system by the use of a weak dc electric field (about 10 to 100 V cm^{-1}), about whose axis they precess with an angle $\bar{\theta} = \arccos\langle \cos\theta \rangle$. This is illustrated in Fig. 3.13, which is intended to represent the case of CH_3I, oriented and with a typical precession angle (by R. J. Beuhler and R. B. Bernstein). Experiments on the reactive scattering of oriented molecules will be discussed in Chapter 6.

A review of the subject has been published (BRO76).

We have now given cursory consideration to the *pre*collision aspects of molecular beams. In the next chapter, we shall briefly outline the *post*collision analysis, i.e., the determination of the attributes of the scattered molecules following collision.

Review literature

Molecular beams
RAM56, HER62, ROS62, FIT63, BER64, HAS64, MCD64, MCD64x, AND65, HER65, PAU65, ZOR65, AMD66, AND66, GRE66, HER66, MUS66, ROS66, STE66, PAU68, STE68, TOE68, SCH69, ROS70, SCH70, MAS71, KIN72, LEE72, FAR73, FLU73, HER73, HER73a, PAR73, PAU73, POL73, VAN73, WON73, EYR74, FAR74, LEV74, TOE74, TOE74a, ZAR74, EYR75, GRI75, LAW75, PAU75, REU75, BRO76, HER76, TOE76, BRO77, FAR77, FAU77, HER77, BER78, ZEW78, BER79, DAV79, FAR79, GRI79, HER79, KWE79, LEV79x, LOS79

Lasers
MAI69, DEM71, FEL71, SIE71, KOM73, KOM73a, LEN74, MOO74, SAR74, STE74, WIL74, BER75, STE75, YAR75, COH76, GRO76, HIN76, JAC76, MOO76, PIM76, STE76, SVE76, WAL76, DEM77, EWI77, GRA77, HAR77, LET77, MOO77, SCH77, SCH77x, SHE77, WIL77, GRU78, OKA78, RED78, STE78, TUR78, ZEW78, BLO79, KOM79, KOM79a, MOO79, RED79, RHO79, ROC79, STI79, SUD79, WAL79, ZAR79, JOH80, KNE80, LEE80, LET80, PHY80, ZAR80, BEN81, CAN81, HAL81, HOU81, JOH81, WAL81, WEI81, WOO81

4 Molecular collisions via crossed beam techniques. Part II: Postcollision product state analysis

We shall now have a look at some of the techniques used to study the *post*collision attributes of the scattered molecules following reactive collision. Two aspects of the subject will be considered. First, how can we measure the products' *internal* state distribution, i.e., $P(v', j')$ and thus $P(E'_{int})$, or better, the state-to-state rate constants $k_{fi}(T)$ or reaction cross sections $\sigma_{fi}(E_{tr})$? Second, how can we determine the products' recoil *velocity vector distribution*, i.e., the product flux (velocity-angle) contour map, from which the detailed differential centre-of-mass (c.m.) reaction cross section can be obtained?

4.1. Products' energy distribution

It is noted that integration over all angles yields the total reaction cross section for given products' translational energy and thus $P(E'_{tr})$, i.e., their relative translational energy distribution [which is only trivially different from $P(E'_{int})$, via total energy conservation]. We shall see examples of the construction of $P(E'_{tr})$ from direct measurements of $P(E'_{vib})$ and $P(E'_{rot})$, which combine to give $P(E'_{int})$ and thus $P(E'_{tr})$. But a given observed $P(E'_{tr})$ can yield only $P(E'_{int})$ and not the separate rotational and vibrational distributions in general; in special favourable cases where the successive vibrational 'bands' in the translational spectrum overlap only slightly, one can separate out the rotational envelopes and thus deduce both $P(E'_{rot})$ and $P(E'_{vib})$. More on this later.

Two methods have been widely used for product internal state analysis, one based on chemiluminescence and the other on laser-induced fluorescence measurements on the nascent products.

Chemiluminescence observations in the visible and UV regions have long been used to ascertain relative vibrational (and even rotational) populations of various electronically excited product molecules (from flames, shock-wave induced reactions, discharges, etc.). Of course, one always has the problem of collisional relaxation to deal with in interpreting the relative populations derived from analysis of the emission spectrum. Because of the rapid rotational relaxation processes, it is difficult to obtain the nascent *rotational* population distribution from any experiment carried out at pressures of more than a few millitorr. Nevertheless, valuable information on electronic-state branching ratios and *vibrational* distributions within a given electronic state of the product is obtained. The analysis of the spectrum requires a knowledge of the line strengths (oscillator strengths for the various observed transitions) and an extensive computer simulation program that attempts to recover the observed, usually overlapping, band structure of the emission spectrum (e.g., by C. Ottinger, D. W. Setser, and others). Of course, the above refers to reactions which are energetic enough to produce electronically excited products. What about reactions that proceed on an adiabatic, ground-electronic-state potential energy surface? These include the familiar ones, referred to many times in these lectures,

$$Cl + HI \rightarrow HCl^{\dagger} + I \qquad (4.1a)$$

$$F + H_2 \rightarrow HF^{\dagger} + H \qquad (4.1b)$$

which produce vibrational population inversions (and thus IR chemiluminescence) but essentially no electronic excitation of products. [It may be noted that the formation of $I^*(^2P_{1/2})$, though energetically possible, has been found to be of negligible importance, and, anyway, our concern now is for the detection of IR photons, regardless of their source.]

4.2. Infrared chemiluminescence for product state analysis

The IR chemiluminescence method was pioneered by J. C. Polanyi and has been well reviewed in the literature, and so we need not elaborate on it here. Suffice it to say that the steady-state IR radiation emitted from the reaction zone (a typical fast-flow system or a 'crossed jet' arrangement is used) is dispersed by a monochromator and its spectrum recorded (Fig. 4.1). Today, a Fourier transform IR spectrometer is used because of its enhanced sensitivity and speed. The emission spectrum is then analyzed, i.e., computer simulated (using known Franck–Condon factors), to yield relative populations of the emitting rovibrational states. Of course, no information is obtained on molecules formed in the ground vibrational states, for the obvious reason that they are 'dark.'

Examples of the results of IR chemiluminescence studies of nascent product state distributions were presented in Chapter 2 (Figs. 2.5 through 2.7). Figure 4.2 (upper) shows an early chemiluminescence spectrum for the HCl from Reaction 4.1a. Figure 4.2 (lower) is a first-overtone spectrum of DCl from the isotopic analogue reaction. The results (such as those shown in Fig. 2.6) can be succinctly presented in the form of a so-called triangle plot, displaying the pattern of energy disposal into product vibration, rotation, and translation. One such plot, that for Reaction 4.1b, is shown in Fig. 4.3.

The most straightforward procedure for going from an observed, well-resolved spectrum such as that of Fig. 4.2 to a table of relative populations of the upper (emitting) states is as follows. First the relative areas under each of the (properly assigned) peaks are measured. Then they must be corrected for

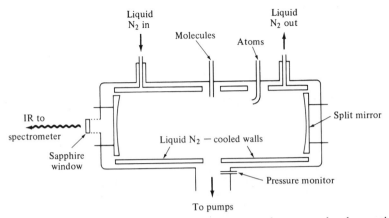

FIG. 4.1. Sketch of chemiluminescence apparatus to study atom-molecule reactions. Adapted from D. H. Maylotte, J. C. Polanyi, and K. B. Woodall, *J. Chem. Phys.* **57**, 1547 (1972); details therein.

all known instrumental discrimination effects. The corrected intensities $I(v'J' \to v''J'')$ are converted to state populations via the expression

$$N_{v'J'} \propto g_{J'} I(v'J' \to v''J'') / [v^4 | R_{v'v''}|^2 S_{J'} F(v'J',v''J'')]$$

where $g_{J'} = 2J' + 1$ is the rotational degeneracy of the emitting states, I the intensity of the transition at frequency v, R the matrix element for the pure vibrational transition, $S = J'$ for P-lines ($J' + 1$ for R-lines), and F the vibration-rotation factor. For typical simple diatomics, such as the products of many of the chemiluminescent reactions, R and F values are usually known (i.e., calculated *ab initio* from the known internuclear potential curves of the diatomic).

For the case in which the resolution of rotational transitions is incomplete, it is customary to assume some trial form of a rotational population distribution and use it to computer-simulate the emission spectrum, varying the shape of the assumed distribution until good agreement with the observed rotational envelope or vibrational band shape is achieved.

4.3. Laser-induced fluorescence for product state analysis

An alternative approach applies lasers to the problem of internal state analysis of nascent products of elementary gas-phase reactions. This is the laser-induced fluorescence (LIF) technique, introduced by R. N. Zare and co-workers, which has proved to be of great utility in the field of reactive molecular scattering. It has the advantage that it is applicable to the analysis of 'dark'

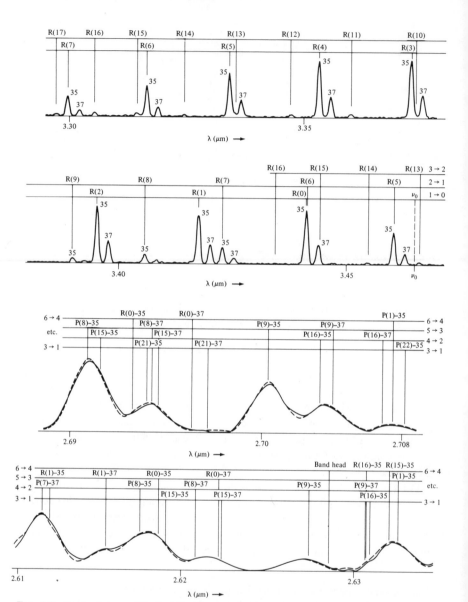

FIG. 4.2. Early data on IR chemiluminescence: (*upper*) nascent HCl molecules from Cl + HI reaction; (*lower*) DCl molecules from Cl + DI. Adapted from Maylotte et al., reference from Fig. 4.1; see also J. C. Polanyi, *Chem. Brit.* **2**, 151 (1966).

products of the reaction, i.e., not restricted to excited (spontaneously fluorescing) product states.

The method of the LIF technique is to irradiate the molecules to be probed with a tunable laser and observe the total (undispersed) fluorescence as the laser-excited states return to the ground state. The lifetimes of these states must be short, however, e.g., $\lesssim 1$ μs. Figure 4.4 shows a schematic apparatus diagram and Fig. 4.5 the LIF excitation spectra of BaBr and BaI produced, respectively, via the reactions:

$$\text{HBr, HI} + \text{Ba} \rightarrow \text{BaBr, BaI} + \text{H} \qquad (4.2)$$

i.e., the fluorescence intensity versus laser wavelength. The BaX products formed in the various v', J' states of the ground electronic state ($X^2\Sigma^+$) are excited to the $C^2\Pi_{3/2}$ state and fluoresce back to the ground state. The $C \rightarrow X$,

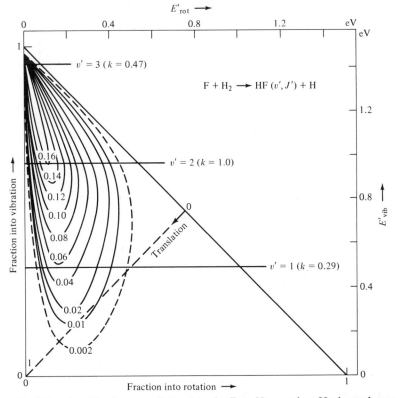

FIG. 4.3. Triangle plot of energy disposal in the $F + H_2$ reaction. Horizontal cuts at $v' = 1$, 2, and 3 indicate HF rotational energy distributions; relative rate constants for each v' obtained by summing over all rotational states with given v'. Adapted from J. C. Polanyi and D. C. Tardy, *J. Chem. Phys.* **51**, 5717 (1969).

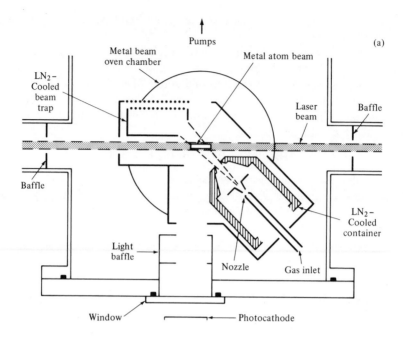

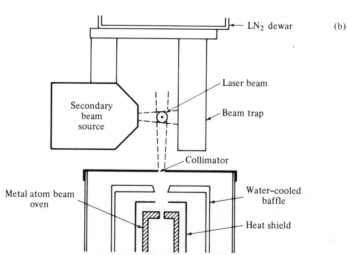

Fig. 4.4. Schematic of molecular beam laser-induced fluorescence apparatus for the detection of nascent products of reactions of alkaline earth atoms with hydrogen halides. Adapted from H. W. Cruse, P. J. Dagdigian, and R. N. Zare, FAR73, p. 277; R. N. Zare, FAR79, p. 7; details therein.

HBr + Ba ⟶ BaBr (X) + H

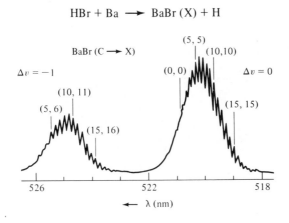

HI + Ba ⟶ BaI (X) + H

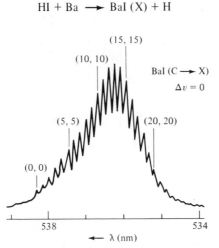

FIG. 4.5. Laser-induced fluorescence from nascent products of HBr, HI + Ba reactions, from references of Fig. 4.4.

$\Delta v = 0$ and -1 sequences are observed for BaBr; however, for the BaI product only the $\Delta v = 0$ sequence is detected. Because of the small rotational (B_v) constants for these molecules, individual product rotational states could not be resolved.

Figure 4.6 is an oversimplified schematic drawing illustrating the essential features of the LIF process for the BaX molecules. The potential energy curve for the upper electronic state (C) lies more or less directly over that for the

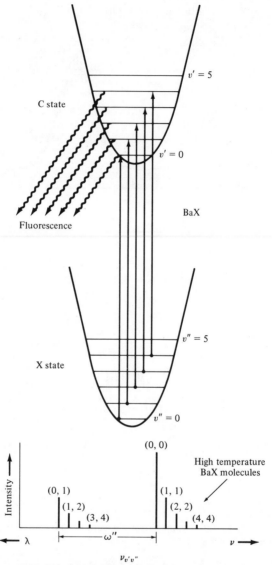

FIG. 4.6. Sketch illustrating principle of laser-induced fluorescence. Example is for hot BaX molecules (X $\equiv$ halogen). See text.

ground (X) state; the upper state's vibrational spacings, ω', are greater than those for the ground state, ω''. For simplicity, we have assumed harmonic oscillator levels so that excitation frequencies (wavenumbers) are given by

$$\nu_{v'v''} = \nu_{00} + v'\omega' - v''\omega'' \tag{4.3a}$$

Shown in Fig. 4.5 are $\Delta v = 0$ excitation transitions, i.e., those for which $v'' = v'$, so that

$$\nu_{v'v''} = \nu_{00} + v'(\omega' - \omega'') \tag{4.3b}$$

Also observed (for BaBr) are $\Delta v = -1$ transitions, for which $v' = v'' - 1$, so that

$$\nu_{v'v''} = \nu_{00} + v'(\omega' - \omega'') - \omega'' \tag{4.3c}$$

Shown at the bottom of Fig. 4.6 is a schematic drawing of the LIF spectrum expected for a 'high-temperature' vapor of BaBr, showing the band origin ν_{00}, labelled (0,0), and the remainder of the $\Delta v = 0$ series at higher frequencies, with constant spacings $\omega' - \omega''$ (from Eqn. 4.3b) and decreasing relative intensities. The $\Delta v = -1$ series is also shown, at longer wavelengths, with the (0,1) line shifted from the (0,0) position by the ground-state vibrational frequency ω'' (Eqn. 4.3c). It should be noted that the fluorescence intensity plotted is the total undispersed fluorescence, i.e., the sum of the intensities of the downward transitions out of the given v' state to the various final v'' states.

The observation of the relative intensities $I_{v'v''}$ of these vibrational bands leads to a set of nascent vibrational populations $N_{v''}$ for the X-state BaX molecules.

One can use as an approximation

$$N_{v''} = I_{v'v''} \bigg/ \left[q_{v'v''}\rho(\nu_{v'v''}) \sum_v \nu_{v'v}^4 q_{v'v} \right] \tag{4.4}$$

where $\rho(v)$ is the laser power density, $q_{v'v''}$ the Franck–Condon factor for the v'',v' transition, and the sum Σ_v is over all vibrational states v of the ground electronic state to which the particular v' (C) state fluoresces. The limitation of Eqn. 4.4 is that it requires Franck–Condon factors for each of the important transitions, implying a good spectroscopic knowledge of the excited (and ground) electronic state. For the BaX molecules, the $C \rightarrow X$, $\Delta v = 0$ system is the strongest, since $B'_v \approx B''_v$. Also, the q values vary by less than a factor of 2 over the range of states excited. The fluorescence sum consists mainly of a single term, namely, that for $\Delta v = 0$, i.e., $q_{v'v''} \approx \delta_{v'v''}$. Thus, under favourable conditions, Eqn. 4.4 reduces simply to a linear proportionality:

$$N_{v'v''} \propto I_{v'v''} \tag{4.5}$$

Thus, the relative intensities of the observed LIF spectrum display directly the relative vibrational populations of the X-state molecules (here, the nascent products, BaX, of the Ba + HX reactions).

Many further developments of the LIF state analysis and detection technique have followed, e.g., by P. J. Dagdigian, R. N. Zare, and others, and its sensitivity has been enhanced sufficiently to make possible population measurements as a function of scattering angle. Recently, J. L. Kinsey, D. E. Pritchard, and K. Bergmann and their respective co-workers have made use of the Doppler shift of the LIF signal to determine the velocity angle distributions of specific rovibrational states of scattered molecules.

There are still more recent developments in laser applications to internal state analysis. One of these is the state-selective technique of laser *multiphoton ionization,* now becoming a practical possibility, largely as a result of the work of P. M. Johnson, F. W. Dalby, V. S. Letokhov, J. Los, E. W. Rothe, R. N. Zare, K. Welge, E. W. Schlag, E. Schumacher, M. A. El-Sayed, L. Goodman, R. B. Bernstein, and others. Since this topic will be discussed in Chapter 10, no further mention will be made here.

4.4. Laser-bolometer methods

Another technique for specific state detection of molecular beams is the IR absorption–bolometer detection method developed by G. Scoles and co-workers T. E. Gough and R. E. Miller, in which a tunable IR laser (e.g., a tunable diode laser or a color-centre laser) is used to excite a particular vibrotational transition in an isolated beam molecule, in flight. The excitation is detected by a sensitive bolometer, which responds to the sum of the translational plus internal energy flux of the molecules impinging on it.

Figure 4.7 shows the bolometer response, i.e., its output signal, proportional to the power delivered to it by a beam of CO as a function of the laser frequency, as the laser is tuned through the R(1) resonance. For comparison, the IR absorption spectrum of CO in a gas cell (the same transition) is shown. Notable is the very narrow line-shape of the molecular beam resonance (due to the perpendicular, nearly Doppler-free irradiation condition). It is possible to measure relative populations of vibrotational states of molecules in beams by this relatively direct energy absorption technique.

Another application of the bolometer to the measurement of internal excitation of molecules in beams is that due to G. W. Flynn, R. B. Bernstein, and their co-workers D. R. Coulter, M. I. Lester, F. R. Grabiner, L. M. Casson, and G. B. Spector, in which the bolometer output is used directly as a monitor of total energy flux, and, by subtraction of the known kinetic energy of the beam, its internal energy can be determined (with and without laser excitation).

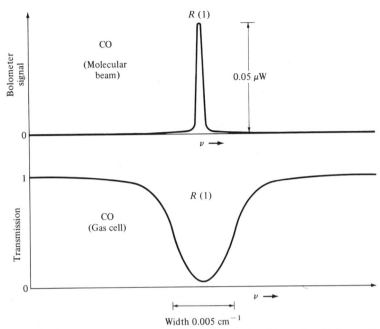

FIG. 4.7. Bolometer detection of vibrational transition in CO molecular beam *(upper)* versus absorption spectrum of CO in the bulk gas. Adapted from T. E. Gough, R. E. Miller, and G. Scoles, *Appl. Phys. Lett.* **30**, 338 (1977); see also *J. Chem. Phys.* **69**, 1588 (1978).

4.5. Products' translational energy distribution measurements

Most of the results on product angular and translational energy distributions have come from more classical crossed molecular beam techniques. Typically, a number-density detector (i.e., a mass spectrometer) or a flux detector (e.g., a surface ionization device) scans the laboratory angle range over which products are found and the velocity distribution of one or more of the products is measured (either by a mechanical velocity analyzer or a so-called time-of-flight chopper).

For determining the reaction cross section, it is preferable to make use of a flux-sensitive detector, i.e., one that responds linearly to the scattered flux or the number of scattered (product) particles striking the detector per unit time, say

$$F = I(\Theta,\Phi)\Omega_d \tag{4.6}$$

Here, $I(\Theta,\Phi)$ is the scattered intensity per unit solid angle Ω at laboratory angle Θ,Φ (i.e., the number of scattered particles per steradian per second directed at the angle Θ, Φ) and Ω_d is the solid angle subtended at the scattering zone by the detector.

An example of such a flux detector is the so-called hot-wire surface ionization detector (usually a narrow ribbon or wire of a high-work-function metal, such as W or Re), which can efficiently ionize alkali atoms or alkali molecules. Nearly every particle striking the detector surface is ionized, the resulting ions are ejected from the hot filament, deflected electrostatically, and counted. The first practical *selective* surface ionization detector was developed by S. Datz and E. H. Taylor and exploited in the first successful observation of a crossed beam reaction, namely,

$$HBr + K \rightarrow KBr + H \tag{4.7}$$

An alternative, more universal detector which is widely used for molecular beam detection is an electron impact or photoionization source in a mass spectrometer. Typically, these are low-efficiency ionizers (e.g., 10^{-6} to 10^{-2} conversion of neutrals to ions). For such detectors the ion current is proportional to the number density n of the neutral particles in the ionizing region, since the probability of ionization is proportional to $n = J/v$, where J is the particle flux density (number per unit area per unit time) and v the laboratory velocity of the particles in the direction of the detector. Thus, to obtain a knowledge of the product flux $F \propto \int_0^\infty vn(v)dv$ and, ultimately, the reaction cross section, one must know the product velocity distribution $n(v)$ at the given scattering angle Ω. Typically, as shown in Fig. 3.1, one uses an electron bombardment ionizer–quadrupole mass spectrometer and records the time-of-flight product distribution at various laboratory scattering angles, usually in the plane of the crossed beams. Most of the recent work in molecular beam scattering makes use of electron impact ionization, mass filter detection, as exemplified by the elegant apparatus of Y. T. Lee and co-workers.

Figure 4.8 shows the earliest observation, by D. R. Herschbach and co-workers G. H. Kwei and J. A. Norris, of a laboratory angular distribution of reactive scattering from crossed beams which led to meaningful information on the shape of a (centre-of-mass) product angular distribution. The reaction is

$$CH_3I + K \rightarrow KI + CH_3 \tag{4.8}$$

and the observations were of the total current K + KI minus that of the K (the nonreactive scattering). Figure 4.9 is a polar plot of the c.m. angular distribution of the KI from this reaction, from a later, detailed study. The strongly anisotropic KI yield, concentrated in the 'backward' hemisphere, implies that a direct, brief encounter at short range suffices to transfer the I atom to the K and form KI, which 'rebounds' energetically backward in the direction from which the K came. (Such 'mechanistic pictures' will be discussed in Chapter

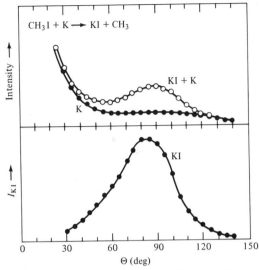

$CH_3I + K \longrightarrow KI + CH_3$

KI + K

K

KI

FIG. 4.8. *Upper:* Laboratory angular distribution of nonreactive scattered K and the sum of that plus the reactive scattered KI from crossed beam study of $CH_3I + K$ system. *Lower:* Difference between these, i.e., the KI product angular distribution. The peak corresponds to backscattering of KI with respect to the incident K direction in the centre-of-mass system. Adapted from D. R. Herschbach, G. H. Kwei, and J. A. Norris, *J. Chem. Phys.* **34**, 1842 (1961).

8.) It is interesting to note that the observed laboratory angular distribution (Fig. 4.8) is quite different from the c.m. distribution (Fig. 4.9). This is because of well-understood kinematic considerations which govern the relationship between the c.m. and laboratory systems.

The relative translational distribution of the nascent products, i.e., the function $P(E'_{tr})$, is most directly determined by measuring the velocity distribution of one or another of these reactively scattered products. Figure 4.10 shows the first example of a velocity analysis (by A. E. Grosser, A. R. Blythe, and R. B. Bernstein), that of the scattering resulting from the collision of a velocity-selected K beam by a thermal beam of HBr. Similar product velocity analyses are required over a wide range of laboratory scattering angles Θ in order to determine the centre-of-mass velocity and angular distribution of the products. The relation between observed laboratory scattering patterns (product flux-velocity-angle distributions) and detailed differential reaction cross sections in the c.m. system was pointed out by T. T. Warnock and R. B. Bernstein (see MRD), but is a little too complicated to go into at this stage. However, it will be worthwhile to consider in some detail the several c.m. cross section functions and their relation to the 'overall' reactive scattering cross section σ_R.

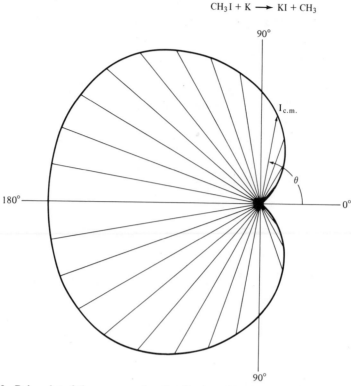

$$CH_3I + K \longrightarrow KI + CH_3$$

FIG. 4.9. Polar plot of the c.m. angular distribution of KI from the $CH_3I + K$ crossed beam reaction. Adapted from R. B. Bernstein and A. M. Rulis, BER73; see also A. M. Rulis and R. B. Bernstein, *J. Chem. Phys.* **57**, 5497 (1972).

4.6. Relation between laboratory and c.m. scattering cross sections

For simplicity, we consider the prototypical atom-molecule exchange reaction:

$$A + BC \rightarrow AB + C \qquad (4.9)$$

A particular exoergic reactive collision viewed in the centre-of-mass system is depicted in Fig. 4.11. Shown is the pair of (antiparallel) incident relative velocity vectors w_A and w_{BC} and one particular pair of outgoing relative velocity vectors for the products, w_{AB} and w_C, scattered with a particular c.m. deflection angle θ. For unpolarized, nonoriented reagents the product flux pattern is cylindrically symmetric around the incident relative velocity axis, i.e., there is no azimuthal angle (ϕ) dependence of the scattered intensity per unit solid angle ($d^2\omega \equiv \sin\theta d\theta d\phi$) c.m. system.

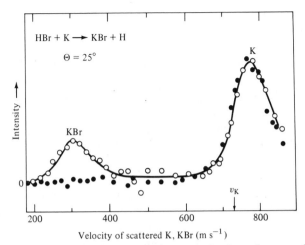

FIG. 4.10. Velocity of reactive and nonreactive scattering from the crossed beam reaction of HBr + K. Velocity-selected K beam (v_K = 732 m s^{-1}), thermal HBr (260 K); laboratory angle Θ = 25° from the K direction. Open circles: detecting K + KBr; solid circles: K only. Difference peak near 300 m s^{-1} is KBr flux. Adapted from A. E. Grosser, A. R. Blythe, and R. B. Bernstein, *J. Chem. Phys.* **42**, 1268 (1965).

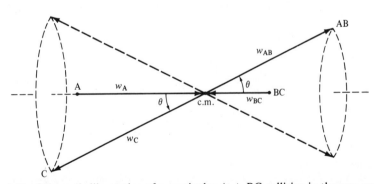

FIG. 4.11. Schematic illustration of a particular A + BC collision in the c.m. system leading to the formation of AB and C, ejected at the indicated scattering angle θ with velocities w_{AB} and w_C, respectively. Because of the axial symmetry (for unpolarized reagents), the products are emitted in cones (with no azimuthal angle dependence). See text.

Each particular collision is characterized by a definite relative translational energy or collision energy in the c.m. system:

$$E_{tr} = \frac{P_A^2}{2m_A} + \frac{P_B^2}{2m_{BC}} = \text{½} \, m_A w_A^2 + \text{½} \, m_{BC} w_{BC}^2 = \text{½} \, \mu_{A,BC} v_{rel}^2 \quad (4.10)$$

where the incident momenta are $P_A = m_A w_A$, $P_{BC} = m_{BC} w_{BC}$, with $P_A + P_{BC} = 0$; $\mu_{A,BC} = m_A m_{BC}/(m_A + m_B + m_C)$ is the reduced mass of the reagents; and $v_{rel} = w_A - w_{BC}$ is the incident relative velocity of the colliding pair. For each particular reactive collision (recall Fig. 2.2), there is a final relative translational energy E_{tr}' such that

$$E_{tr}' = \frac{P_{AB}^2}{2m_{AB}} - \frac{P_C^2}{2m_C} = \text{½} \, m_{AB} w_{AB}^2 + \text{½} \, m_C w_C^2 = \text{½} \, \mu_{AB,C} v_{rel}'^2 \quad (4.11)$$

where the final relative velocity is $v_{rel}' = w_{AB} - w_C$, the final reduced mass is $\mu_{AB,C} = m_{AB} m_C/(m_A + m_B + m_C)$, and $P_{AB} = m_{AB} w_{AB}$, $P_C = m_C w_C$, with $P_{AB} = -P_C$. These equations imply a simple relation between the 'measurable' c.m. recoil velocity of one of the products, e.g., AB, and the final relative translational energy:

$$E_{tr}' = C w_{AB}^2 \quad (4.12)$$

where C is a constant involving masses m_A, m_B, m_C only. Thus the integral reaction cross section can be written

$$\sigma_R = \int_0^E dE' \, \frac{d\sigma(E')}{dE'}; \quad 0 \le E' \le E \quad (4.13)$$

where E is the total available energy (Fig. 2.2) and the subscript 'tr' has been dropped.

Two types of differential cross sections stem from a common 'detailed differential cross section':

$$\frac{d^2\sigma(\theta)}{d^2\omega} = \int_0^E dE' \, \frac{d^3\sigma(\theta, E')}{d^2\omega dE'} \quad (4.14)$$

$$\frac{d\sigma}{dE'} = 2\pi \int_0^\pi d\theta \, \sin\theta \, \frac{d^3\sigma(\theta, E')}{d^2\omega dE'} \quad (4.15)$$

Detailed differential cross sections are usually deduced from measurements of the product flux-velocity-angle distributions in the laboratory system (first reported by R. B. Bernstein and co-workers A. E. Grosser, T. T. Warnock, and K. T. Gillen and by J. H. Birely and D. R. Herschbach). An early example is that shown (in the form of a polar flux contour map) in Fig. 4.12 for the KI from the reaction

$$K + I_2 \rightarrow KI + I \quad (4.16)$$

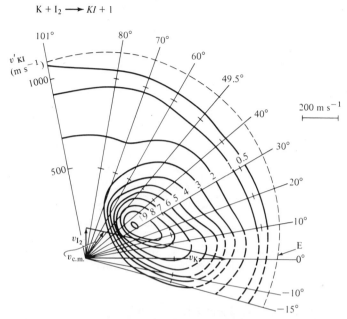

FIG. 4.12. Laboratory flux-velocity distribution (contour map superimposed upon velocity vector triangle) for the KI from the K + I$_2$ reaction. Outer dashed circle corresponds to energy conservation limit. Adapted from K. T. Gillen, A. M. Rulis, and R. B. Bernstein, *J. Chem. Phys.* **54**, 2831 (1971); details therein.

Such data are then transformed from the laboratory to the c.m. system (using relations developed by T. T. Warnock and R. B. Bernstein) to produce the detailed differential cross sections in c.m. velocity-space, i.e., a quantity proportional to $d^3\sigma(\theta, w')/d^2\omega dw'$, where w' is the magnitude of the c.m. recoil velocity of the detected product (e.g., w_{KI} in Reaction 4.16), displayed in Fig. 4.13. Such c.m. flux-velocity-angle contour maps provide both quantitative and qualitative information about the reaction dynamics, as will be discussed in Chapter 8.

It is noted that

$$\frac{d^3\sigma(\theta, w')}{d^2\omega dw'} = \frac{d^3\sigma(\theta, E')}{d^2\omega dE'} \times \frac{dE'}{dw'} \tag{4.17}$$

where dE'/dw' is obtained from Eqn. 4.12, so that

$$\frac{d^3\sigma(\theta, E')}{d^2\omega dE'} \propto \frac{1}{w'} \frac{d^3\sigma(\theta, w')}{d^2\omega dw'} \tag{4.18}$$

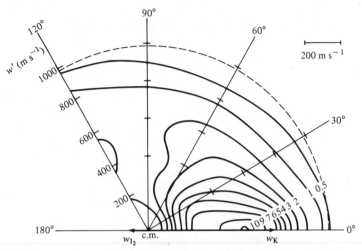

FIG. 4.13. Centre-of-mass distribution of KI from $K + I_2$ reaction; based on results on Fig. 4.12; outer dashed circle corresponds to conservation limit. See reference cited in Fig. 4.12 for details.

The analogues of Eqns. 4.14 and 4.15 are thus

$$\frac{d^2\sigma(\theta)}{d^2\omega} = \int_0^{w'_{max}} dw' \frac{d^3\sigma(\theta,w')}{d^2\omega dw'} \tag{4.19}$$

where the maximum recoil velocity of the (detected) product is $w'_{max} = (E/C)^{1/2}$ and

$$\frac{d\sigma(w')}{dw'} = 2\pi \int_0^{\pi} d\theta \sin\theta \frac{d^3\sigma(\theta,w')}{d^2\omega dw'} \tag{4.20}$$

Various shorthand notations for some of these cross section functions are often used, such as

$$I(\theta,w') \text{ and } I(\theta,E') \text{ for } \frac{d^3\sigma(\theta,w')}{d^2\omega dw'} \text{ and } \frac{d^3\sigma(\theta,E')}{d^2\omega dE'}$$

respectively, and $I(\theta)$ or $P_\theta(\theta)$ for $d^2\sigma(\theta)/d^2\omega$, sometimes normalized such that

$$2\pi \int_0^{\pi} d\theta \sin\theta P(\theta) = 1 \tag{4.21}$$

and $P_{w'}(w')$ or $P_{E'}(E')$ for $d\sigma(w')/dw'$ and $d\sigma(E')/dE'$, respectively, sometimes normalized:

$$1 = \int_0^{w'_{max}} dw' \, P_{w'}(w') = \int_0^{E} dE' \, P_{E'}(E') \tag{4.22}$$

Averaged values are sometimes reported, e.g.,

$$\overline{w'} = \int_0^{w'_{max}} dw'\, w'\, P_{w'}(w')$$ (4.23)

or

$$\overline{E'} = \int_0^E dE'\, E'\, P_{E'}(E')$$ (4.24)

or an average fraction such as that of the available energy disposed into products' translation:

$$\overline{f}_{T'} \equiv \overline{E'}/E ; \quad 0 \le \overline{f}_{T'} \le 1$$ (4.25)

Thus

$$\overline{f}_{V'} = 1 - \overline{f}_{T'}$$ (4.26)

is the average fraction of the available energy partitioned into the products' internal energy.

4.7. Products' translational energy distributions

Figure 4.14 shows a typical plot of $P(w')$ for the KI product of the crossed molecular beam reaction shown in Eqn. 4.8. Figure 4.15 is the corresponding plot of $P(E')$ for the same reaction.

The lack of vibrational structure in such plots (compared with the envelopes of Fig. 2.7, for example) is due to the unavoidable spread in incident relative velocities and the resulting smearing of the products' translational energies, as

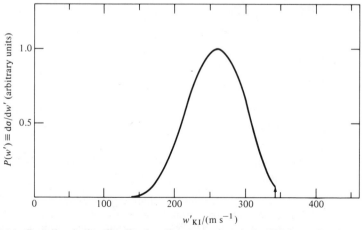

FIG. 4.14. Recoil velocity distribution (in c.m. system) for KI from the $CH_3I + K$ reactions; arrow denotes conservation limit. See references of Fig. 4.9 for details.

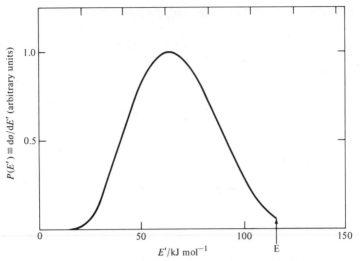

FIG. 4.15. Products' relative translational energy distribution for the CH₃I + K reaction; arrow denotes conservation limit. Based on results of Fig. 4.14.

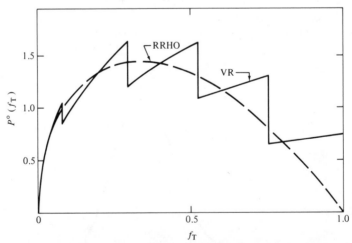

FIG. 4.16. Calculated translational energy distribution (f_T is the fraction of the available energy going into products' translation) for statistical, or 'prior expectation,' product internal state distribution. Dashed curve: rigid-rotor harmonic oscillater (RRHO) model; solid curve: vibrating rotor (VR) model for the Cl + HI reaction. Adapted from R. D. Levine and R. B. Bernstein, LEV73.

74

well as the finite resolution of the velocity analyzer. Figure 4.16 shows a calculated translational recoil distribution (assuming a statistical internal state distribution) for a realistic atom-molecule system, assuming monochromatic beams and perfect velocity resolution. It is clear that product state resolution via translational energy analysis is inherently difficult except in special, favourable circumstances (e.g., sparsely populated, well-separated levels).

Since it is the purpose of this chapter not to discuss results but rather to emphasize methods, we shall not deal further with the subject of velocity-angle distributions. In Chapters 6 through 9 we shall review in some detail a body of reactive scattering results and present ample discussion thereof. Suffice it to say at this point that reactive molecular beam scattering measurements (e.g., velocity analysis) are complementary to and may be correlated with data on product state analysis carried out by chemiluminescence or laser-induced fluorescence methods.

4.8. Recent advances in product state analysis

Recent technical developments in the LIF field have made it the method of choice for many systems: in certain favourable cases (as mentioned in Section 4.3) it has been possible to measure the angular distribution of $P(v',J')$ and even (via Doppler shift) the laboratory velocity distribution of the particular product state!

In addition, the polarization of the rotating product molecules with respect to the laboratory frame has been measured in several instances by several different methods. The first observations of the rotational angular momentum polarization of nascent diatomic molecules were by D. R. Herschbach et al. using the electric deflection technique. Later, J. P. Toennies et al. measured the J, M_J distribution of such product molecules by means of the electric quadrupole focusing technique. Direct polarization analysis of product chemiluminescence and laser-induced fluorescence has also been carried out, by J. P. Simons et al. and R. N. Zare and co-workers. The theory relating product rotational angular momentum polarization to the molecular dynamics has been developed in detail by D. R. Herschbach and by R. N. Zare and their co-workers, and classical trajectory studies reported by J. C. Polanyi et al. Experimental results and interpretation will be discussed in Chapter 6.

Review literature

CAR72, BER73, CRU73, PAR73, MOO74, STE74, TOE74, TOE74a, WIL74, ZAR74, MCG75, TOE76, FAR77, FAU77, KIN77, MOO77, PAO77, ZEW78, BER79, MOO79, MOO79a, ZAR79, CRO80, KNE80, LET80, PHY80, SMI80, ZAR80, BAR81, BEN81, HAA81, JOR81, JOR81a, MCG81, MOO81, REI81, RON81

5 Elastic and inelastic scattering as a manifestation of intermolecular forces

5.1. Nonreactive collisions

Before we can deal properly with the concept of the reactive collision, we must back up a little and consider nonreactive encounters of atoms and molecules, which are simpler to understand and to describe in quantitative terms. The goal: to predict from first principles the observable elastic and inelastic scattering behaviour of simple molecules. It will be seen that this task divides naturally into two parts: first, that of determining the intermolecular forces, and second, that of calculating the dynamics of the collisions, governed by these forces. Anticipating the remainder of this chapter, it turns out that the experimental scattering data are usually used to deduce the forces and not the other way around! Nevertheless, the results are found to be self-consistent and satisfying, and lead to confidence in our ability to deal with nonreactive molecular collisions.

Given a quantitative knowledge of the intermolecular forces, i.e., the interaction potential for the molecular system, it is now possible to calculate (i.e., 'predict') all the observable scattering behaviour, elastic as well as inelastic, of the molecules undergoing collisions. One can then go further and predict the equilibrium and transport properties of the gaseous molecules up to moderate densities. However, for most systems *ab initio* calculated intermolecular forces are not accurately known, and so much of the emphasis has been on 'inverting' experimental transport, gas imperfection, and (more recently) molecular beam scattering data to ascertain the interaction potential. The landmark work in the field is the classic by J. O. Hirschfelder, C. F. Curtiss, and R. B. Bird, HIR54. However, in the last 20 years molecular beam scattering studies have revolutionized the field, and there now exist accurately determined intermolecular potentials for a variety of atom-atom, atom-molecule, and molecule-molecule systems, thanks to pioneering work from the laboratories of H. Pauly, R. B. Bernstein, E. F. Greene, J. Reuss, G. Scoles, A. Kuppermann, and especially Y. T. Lee.

5.2. Elastic scattering

From elastic scattering measurements (angular distributions and total cross sections as a function of the collision energy), it is possible to deduce the orientation-averaged interparticle potential function. From inelastic scattering

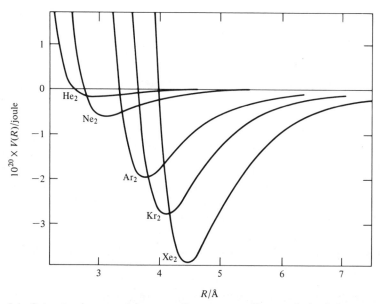

FIG. 5.1. Interatomic potential curves for rare-gas dimers, derived from crossed atomic beam measurements of the angular distribution of elastic scattering over a range of collision energies. Adapted from J. M. Farrar, T. P. Schafer, and Y. T. Lee in *Transport Phenomena* (J. Kestin, ed.), A.I.P. Conference Proceedings No. 11 (1973).

data (e.g., rotational and vibrational excitation/de-excitation cross sections), information on the anisotropic (as well as isotropic) aspects of the potential can be ascertained. In the few cases where reliable *ab initio* computations of the intermolecular potential surface are available, they accord well with the experimentally deduced potentials, thus adding confidence to the inversion procedures.

For the simple case of the interaction of closed-shell atoms, e.g., Ar-Ne, or of one closed-shell atom with an S-state atom, e.g., Ar-K, the ground electronic state interaction potential is of a simple 'Lennard–Jones-like' (LJ) form, as shown in Fig. 5.1 for rare-gas dimers. The potential is strongly repulsive at small separations R, with an attractive well corresponding to the existence of a weakly bound van der Waals diatomic molecule, or 'dimer,' and a long-range attractive 'tail' of the form $V(R) \sim - C_s R^{-s}$ ($s \geq 2$), asymptotically approaching zero interaction energy. For the closed-shell case, $s = 6$. The simplest functional forms that can fairly well represent these interactions are of the LJ (n,6) type, e.g.,

$$V^*(R^*) = \left(\frac{6}{n-6}\right)(R^*)^{-n} - \left(\frac{n}{n-6}\right)(R^*)^{-6} \qquad (5.1)$$

or the exponential $(\alpha,6)$ form, e.g.,

$$V^*(R^*) = \left(\frac{6}{\alpha - 6}\right) e^{-\alpha(R^*-1)} - \left(\frac{\alpha}{\alpha - 6}\right)(R^*)^{-6} \qquad (5.2)$$

where the 'reduced variables,' indicated by asterisks, are used as follows:

$$V^* \equiv V/\epsilon \qquad (5.3)$$
$$R^* \equiv R/R_e$$

where ϵ is the depth of the attractive well and R_e the internuclear separation at the minimum of the well.

Detailed beam scattering studies at relatively low collision energies have revealed the shape of the potential wells for many van der Waals dimers (confirmed by spectroscopic measurements of the vibrational levels of these weakly bound molecules, first carried out by Y. Tanaka and others). Scattering measurements at high collision energies (a field pioneered by I. Amdur) have probed the repulsive cores of the interaction potential for many systems, including molecules as well as atoms. (An 'effective' spherically symmetric potential is generally assumed to approximate the actual intermolecular interaction potential surface.) The scattering of polar molecules by other polar molecules gives information on long-range dipole-dipole forces, as first shown by H. Schumacher, R. B. Bernstein, and E. W. Rothe and extended by R. J. Cross, R. G. Gordon, and D. R. Herschbach.

Before getting into the details of the elastic scattering process, it would be well to lay the groundwork for the understanding of the basis of rotationally and vibrationally inelastic processes. For simplicity let us consider atom-molecule collisions. An impact can alter the rotational and vibrational state of a molecule, resulting in a transfer of energy between the internal and external degrees of freedom of the system. For example, a fast collision can induce rotational (and, with lower efficiency, vibrational) excitation in the molecule; even a slow encounter can suffice to de-excite rotationally or vibrationally excited molecules and transform their internal energy into the relative translational energy of the departing collision partners. Collisions directed 'off the axis' of the molecule are especially effective in altering its rotational state, whereas impulsive collisions directed along a bond axis are most efficient in promoting vibrational excitation/de-excitation.

Let us begin with the simplest process, elastic atom-molecule scattering, by considering the interaction of a spherically symmetric, ground-state atom (e.g., a rare-gas or S-state atom) with a closed-shell, rigid diatomic molecule, as depicted in Fig. 5.2. The lowest electronic potential energy surface can be expressed in the form of radial and angular parts, via the Legendre expansion:

$$V(R,\gamma) = \sum_{\lambda} v_\lambda(R) P_\lambda(\cos\gamma) \qquad (\lambda = 0,1,2,\dots) \qquad (5.4)$$

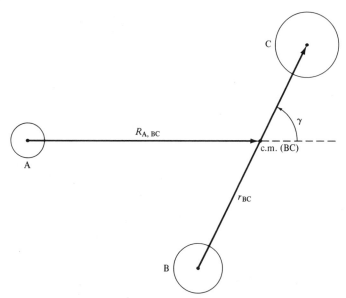

FIG. 5.2. Coordinates to describe A + BC interaction. $R_{A,BC}$ is the vector between A and the c.m. of BC; r_{BC} is the vector corresponding to the internuclear separation of BC; $\gamma \equiv \arccos(\boldsymbol{R} \cdot \boldsymbol{r})$ is the indicated angle of orientation of the molecular axis.

where R is the magnitude of $\boldsymbol{R}_{A,BC}$ (i.e., the separation between the atom A and the centre of mass of the diatom BC) and γ the angle between $\boldsymbol{R}_{A,BC}$ and $\boldsymbol{r}_{BC}$ (the internuclear separation vector of the diatom). For the discussion of elastic and rotationally inelastic scattering, it is convenient to assume that the diatomic molecule is a rigid rotor, i.e., $r_{BC} = (r_e)_{BC}$, a constant. The orientation-averaged potential energy depends only upon R and is evaluated simply:

$$\overline{V}(R) \equiv \langle V(R) \rangle_\gamma \equiv \int_0^\pi V(R,\gamma) 2\pi \sin\gamma \, d\gamma = v_0(R) \qquad (5.5)$$

from the orthonormality of the Legendre functions. Typical radial functions $v_0(R)$, $v_1(R)$, etc., are shown in Fig. 5.3. The orientation-averaged potential curve is of the familiar form, strongly repulsive at small separations with an attractive well corresponding to the van der Waals adduct $(A \cdots BC)$ and a long-range attractive 'tail' asymptotically approaching zero interaction energy.

   An alternative presentation of a potential surface such as that of Eqn. 5.4 is to evaluate cuts of $V(R)$ at various fixed values of γ, as shown in the upper portion of Fig. 5.4. Still another is a polar plot of the equipotential contours, (the lower portion of Fig. 5.4 and Fig. 5.5). For high collision energies, a classical rigid ellipsoid model of the rotational energy transfer process is found to

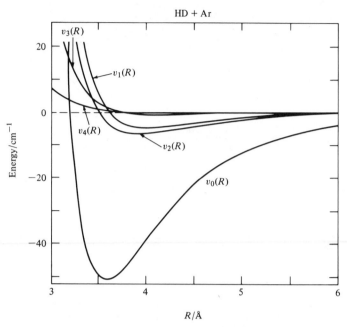

FIG. 5.3. Radial functions $v_0(R)$, ... , $v_4(R)$ describing the HD + Ar interaction potential. Adapted from H. Kreek and R. J. LeRoy, *J. Chem. Phys.* **63**, 338 (1975).

be fairly successful. Both classical and quantal treatments have been used to deal with rotational excitation over a wide energy region, as will be mentioned again later. For the present, let us confine our attention to the elastic scattering from the orientation-averaged potential, i.e., scattering by a central potential, $\overline{V}(R) \equiv V(R)$.

The key feature of elastic scattering is that there is no change in the kinetic energy of the colliding partners as a result of the collision; thus the magnitude of the final relative velocity of the colliding pair is the same as its initial value. However, the *direction* of the relative velocity will, in general, be altered, and so the collision serves to scatter the partners over a range of angles. The angular distribution of the scattering contains information on the interaction potential, since $V(R)$ governs the trajectory of one particle with respect to the other, as will be seen below.

Figure 5.6 shows a typical trajectory for the scattering of a light projectile of mass μ by a stationary target particle, assuming a realistic intermolecular potential for $V(R)$. The initial relative velocity vector $\boldsymbol{v}_r$ is effectively rotated in the course of the collision through an angle θ. This deflection angle depends upon the initial conditions, such as the 'impact parameter' b, the initial value of v_r (and thus the initial relative translational energy $E = \tfrac{1}{2}\mu v_r^2$), and the potential function. A classical mechanical analysis of the elastic collision pro-

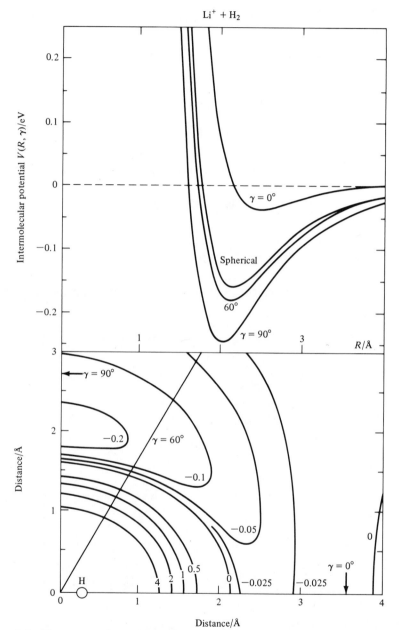

Fig. 5.4. *Upper:* cuts through potential surface for $Li^+ + H_2$ system for different orientation angles γ; *lower:* polar plot of equipotential contours for same (energy in eV). Adapted from R. A. LaBudde and R. B. Bernstein, *J. Chem. Phys.* **55**, 5499 (1971); based on W. A. Lester Jr., *J. Chem. Phys.* **53**, 1511 (1970).

cess is straightforward. Following HIR54, it is easy to show that the deflection angle θ is given by

$$\theta = \pi - 2b \int_{R_0}^{\infty} dR \, R^{-2} \left[1 - \frac{V(R)}{E} - \frac{b^2}{R^2} \right]^{-1/2} \qquad (5.6)$$

where R_0 is the distance of closest approach, defined by the implicit equation

$$1 - [V(R_0)/E] - b^2/R_0^2 = 0 \qquad (5.7)$$

Later we shall discuss the systematics of θ as a function of b, E, and $V(R)$. For the present, we are concerned with its observational implications.

5.3. Laboratory measurements of elastic scattering

Let us therefore return to the laboratory system and see how a crossed molecular beam arrangement makes it possible to study the deflections and thus the angular distribution of the elastic scattering, governed by $V(R)$. For simplicity assume two mono-velocity beams (particles of Types 1 and 2, respectively) crossed at a convenient angle, as shown in Fig. 5.7. (Frequently the incident

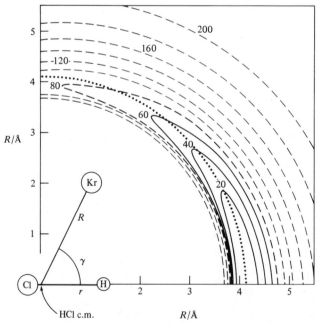

FIG. 5.5. Polar plot of equipotential contours for the Kr···HCl system (energy in cm^{-1} relative to the minimum at $\gamma = 0$). Adapted from J. M. Hutson, A. E. Barton, P. R. Langridge-Smith, and B. J. Howard, *Chem. Phys. Lett.* **73**, 218 (1980).

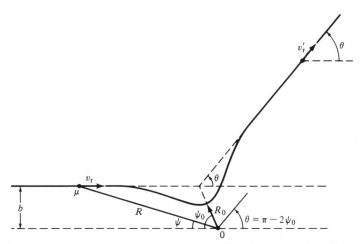

FIG. 5.6. A trajectory in the c.m. system corresponding to an initial relative velocity v_r and impact parameter b for a reduced mass μ. The final deflection angle is $\theta = \pi - 2\psi_0$, where ψ_0 is the indicated angle of the apse. For details, see MRD, Section 2.3.4.

angle is chosen to be 90°.) Although the relative velocity v_r is fixed (determined by v_1 and v_2), there is no control at the microscopic level of the impact parameter b, and so all values of b are represented. According to classical mechanics, for a collision with a given b (and a given E), there is a definite scattering angle θ (in the centre-of-mass system) corresponding to the rotation of the relative velocity vector from v_r to v'_r (primes are used to denote postcollision). The result of one such collision is shown in Fig. 5.7, for which the rotation of v_r happens to be in the plane of the crossed beams, the chosen plane for the scattering detector. Thus the scattered molecules reaching the detector set at a 'laboratory angle' Θ with respect to the beam of Type 1 particles correspond to molecules whose scattering angle (in the c.m. system) is θ.

Let us write down a few important kinematic relationships. The first of these is simply the definition of the relative velocity in terms of the laboratory velocities of the two colliding partners:

$$v_r = v_1 - v_2 \tag{5.8a}$$

The final relative velocity is, similarly,

$$v'_r = v'_1 - v'_2 \tag{5.8b}$$

Conservation of total momentum in the course of the collision can be expressed in terms of the velocity of the c.m., v_c:

$$M v_c = m_1 v_1 + m_2 v_2 \quad (M \equiv m_1 + m_2) \tag{5.9a}$$

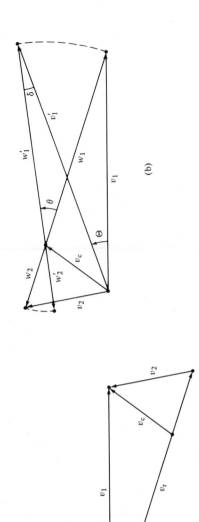

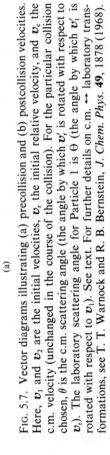

(b)

(a)

FIG. 5.7. Vector diagrams illustrating (a) precollision and (b) postcollision velocities. Here, v_1 and v_2 are the initial velocities, v_r the initial relative velocity, and v_c the c.m. velocity (unchanged in the course of the collision). For the particular collision chosen, θ is the c.m. scattering angle (the angle by which v'_r is rotated with respect to v_r). The laboratory scattering angle for Particle 1 is Θ (the angle by which v'_1 is rotated with respect to v_1). See text. For further details on c.m. ↔ laboratory transformations, see T. T. Warnock and R. B. Bernstein, *J. Chem. Phys.* **49**, 1878 (1968).

so that

$$v_c = (m_1/M)v_1 + (m_2/M)v_2 \qquad (5.9b)$$

a constant throughout the collision. The total kinetic energy (conserved) is given by

$$\text{K.E.} = \tfrac{1}{2}m_1 v_1^2 + \tfrac{1}{2}m_2 v_2^2 = \tfrac{1}{2}m_1(v_1')^2 + \tfrac{1}{2}m_2(v_2')^2 \qquad (5.10a)$$

$$= \tfrac{1}{2}M v_c^2 + \tfrac{1}{2}\mu v_r^2 \qquad (5.10b)$$

where $\mu \equiv m_1 m_2/M$ is the reduced mass of the colliding pair. The first term in Eqn. 5.10b is the kinetic energy of the centroid (c.m.) in the laboratory system; the second term is the relative kinetic energy of the colliding partners, i.e., the collision energy E (the incident relative translational energy, equal to the final relative translational energy, since we are dealing only with elastic collisions).

Next we consider the velocities of each colliding partner with respect to the (moving) centroid, denoted by w_1 and w_2, respectively. Thus

$$v_r = w_1 - w_2 \qquad \text{and} \qquad v_r' = w_1' - w_2' \qquad (5.11)$$

Also, by construction, we have the relations

$$v_1 = v_c + w_1 \qquad \text{and} \qquad v_1' = v_c + w_1' \qquad (5.12a)$$

and

$$v_2 = v_c + w_2 \qquad \text{and} \qquad v_2' = v_c + w_2' \qquad (5.12b)$$

i.e., the laboratory velocity of a given particle is the sum of the velocity of the centroid plus the velocity of the particle relative to the centroid (its c.m. velocity).

Momentum conservation in the c.m. system can be written

$$m_1 w_1 + m_2 w_2 = m_1 w_1' + m_2 w_2' = 0 \qquad (5.13)$$

from which we can 'partition' the relative velocity vectors v_r (and v_r') into their constituent 'components' w_1, w_2 (and w_1', w_2').

Referring back to Fig. 5.7, the laboratory velocity vector v_1' of Particle 1 scattered by the c.m. angle θ, given by Eqn. 5.12a, is directed at an angle Θ from its original direction (v_1). Thus the scattering detector set at an angle Θ with respect to the incident beam will record particles of Type 1 from a collision characterized by a c.m. deflection θ. (Note that all such detected particles have the same laboratory speed, v_1'.) By measuring the flux of molecules scattered into the acceptance cone of the detector at various values of Θ, we can deduce the intensity of scattering at corresponding angles θ in the c.m. system.

Knowing the incident flux of each beam and the beam intersection volume, we can ascertain the so-called differential (solid angle) cross section for elastic scattering at the given E, which can in principle be computed from a knowledge of the potential function $V(R)$. The question of the sensitivity of the differential cross sections to the shape of the potential curve is of considerable importance and will be discussed later on. Suffice it to say for the present that it is at the level of the cross sections that theory can be most directly compared with experiment.

Now we must relate the c.m. differential cross section to the observable scattering. The probability of a particle of Type 1 being scattered by a c.m. angle θ into an element of solid angle $d\omega$ ($\equiv 2\pi \sin\theta d\theta$) per steradian per unit time is

$$P_1 = n_2 v_r \frac{d\sigma(\theta)}{d\omega} \tag{5.14}$$

where $d\sigma(\theta)/d\omega$ is the differential solid angle cross section (the 'collision area' per steradian) and n_2 the number density of particles of Type 2 (i.e., number per unit volume) in the beam intersection volume (ΔV). Then the number of Type 1 particles scattered by an angle θ into $d\omega$ per unit time is

$$d\dot{N}_1(\theta) = (n_1 \Delta V)(P_1 d\omega) = n_1 n_2 v_r \frac{d\sigma(\theta)}{d\omega} \Delta V d\omega \tag{5.15}$$

where n_1 is the number density of Type 1 particles in the crossed beam zone of volume ΔV. Thus the total intensity (i.e., number per unit time) of Type 1 particles scattered into the laboratory solid angle subtended by the detector Ω_d at a laboratory angle Θ corresponding to the c.m. angle θ (Fig. 5.7) is given by

$$I_1(\Theta) \equiv \int_{\Omega_d} \Delta \dot{N}_1(\Theta) = \int_{\Omega_d} \frac{d\dot{N}_1}{d\omega} \frac{d\omega}{d\Omega} d\Omega \approx \frac{d\dot{N}_1}{d\omega} J\Omega_d \tag{5.16}$$

where J is the Jacobian for the solid angle transformation from laboratory to c.m., $J \equiv \left| \dfrac{d\omega}{d\Omega} \right|$. Thus

$$I_1(\Theta) = n_1 n_2 v_r \Delta V \frac{d\sigma(\theta)}{d\omega} J\Omega_d \tag{5.17}$$

The Jacobian can be simply evaluated by inspection of Fig. 5.8: an element of area dA perpendicular to w_1' subtends a solid angle $dA/(w_1')^2$ at the centroid, but the solid angle it subtends at the origin is $d\Omega = (A/\cos\delta)/(v_1')^2$. Thus

$$J \equiv \left| \frac{d\omega}{d\Omega} \right| = (v_1')^2/(w_1')^2 |\cos\delta| \tag{5.18}$$

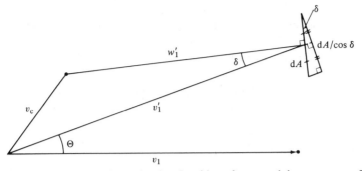

FIG. 5.8. Diagram to assist in evaluating Jacobian of c.m. ↔ laboratory transformation. See reference cited in Fig. 5.7.

From Eqn. 5.17, the intensity of the Type 1 particles detected is then

$$I_1(\Theta) = n_1 n_2 v_r \Delta V \frac{(v_1')^2}{(w_1')^2 |\cos\delta|} \Omega_d \frac{d\sigma(\theta)}{d\omega} \tag{5.19}$$

with an analogous result for Type 2 particles. For a given (elastic) angular distribution experiment at a given collision energy, v_r, n_1, n_2, ΔV, and Ω_d are constant (and $w_1' = w_1$ by energy conservation), so that

$$I(\Theta) \propto \frac{(v_1')^2}{|\cos\delta|} \frac{d\sigma(\theta)}{d\omega} \tag{5.20}$$

where v_1', δ, and θ are given at a specified Θ by the geometric relationships (of purely kinematic origin) of Fig. 5.7. Equation 5.20 allows us to determine the differential solid angle cross section in the c.m. system from laboratory measurements of the angular distribution of beam scattering. (It must be mentioned here that the above brief discussion is much oversimplified. In general, one must take cognizance of experimental velocity spreads and the finite size of beams and detector as well as the possibility of contributions from more than one c.m. angle θ to a given laboratory angle Θ. For further information, see MRD and review chapters.)

5.4. Scattering in the centre-of-mass system

Next we return to the c.m. system and relate the differential cross section to the particle trajectories, governed by the impact parameter b, the collision energy E, and, of course, the interaction potential $V(R)$.

It is shown in HIR54 that the relative motion of the colliding partners in the c.m. system is equivalent to the motion of a projectile particle of mass μ with respect to an infinitely massive (fixed) target, interacting according to the original interparticle potential $V(R)$. Figure 5.9 is a construction intended to serve as the basis of the desired relationships. Assume a uniform incident flux of projectile particles, say j (number per unit area per unit time), with given velocity v_r. Consider the annular ring shown; all projectile particles crossing the annulus with impact parameters within the range between b and $b + db$ will be deflected within an angular range between θ and $\theta + d\theta$, i.e., scattered into the annular θ ring shown. Conservation of flux is expressed

$$2\pi b\,db\,j = 2\pi\sin\theta\,|d\theta|\,\frac{d\sigma(\theta)}{d\omega}\,j \qquad (5.21a)$$

i.e.,

$$d\sigma(\theta) = 2\pi\,b\,db \qquad (5.21b)$$

where the factor $2\pi\sin\theta d\theta$ ($= \int_{\phi=0}^{2\pi}\sin\theta d\theta d\phi$ represents the total solid angle of the θ ring. Thus the differential solid angle cross section is given by

$$\frac{d\sigma(\theta)}{d\omega} = \frac{b}{\sin\theta}\left|\frac{d\theta}{db}\right|^{-1} \qquad (5.22)$$

(the absolute value is used since the dependence of θ upon b typically involves both positive and negative signs and the cross section is intrinsically positive).

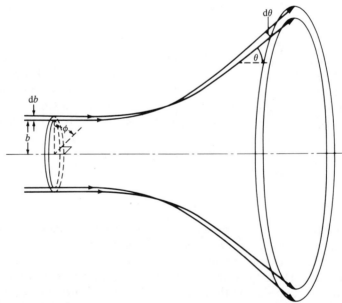

FIG. 5.9. Trajectories for which b lies in the range between b and $b + db$ lead to angular deflections within the range θ to $\theta + d\theta$. For details, see MRD, Section 3.1.3.

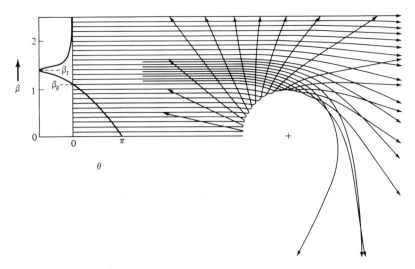

FIG. 5.10. Classical trajectories as a function of the reduced impact parameter β at a given collision energy for a realistic intermolecular potential. The graph on the left shows the dependence of the angle of deflection θ upon β. Adapted from H. Pauly, PAU79.

According to classical mechanics, given $V(R)$, all that is required to calculate the cross section is to evaluate $\theta(b)$ via the quadrature of Eqn. 5.6, then the derivative $d\theta/db$ at the desired θ, and thus $d\sigma(\theta)/d\omega$ via Eqn. 5.22.

Figure 5.10 shows typical 'classical trajectories' for a realistic intermolecular potential (similar to Eqns. 5.1 and 5.2). A 'reduced' impact parameter $\beta \equiv b/R_e$ is used. Note that for $\beta = 0$, a 'head-on' collision, $\theta = \pi$; increasing values of β lead to smaller deflection angles. For $\beta \gg 2$, a 'grazing collision,' $\theta \approx 0$ (negligible deflection). Negative θ values correspond to 'net attractive' trajectories, for the range of b shown. There are two special values of b that are of interest, corresponding to β_g and β_r in Fig. 5.10. For the impact parameter corresponding to β_g the net deflection is zero (though the trajectory shows attraction initially, then repulsion). This is termed the 'glory impact parameter.' For $\beta = \beta_r$, the deflection angle has assumed its greatest negative value, characteristic of the 'most attractive' trajectory. This is called the 'rainbow impact parameter' (by analogy to optics). Obviously, both β_g and β_r depend upon the collision energy and the potential function.

From the qualitative dependence of θ upon b, the main features of the 'classical' angular distribution function, $d\sigma(\theta)/d\omega$, can be immediately predicted (via Eqn. 5.22). It is instructive to compare the results for scattering by a realistic intermolecular potential with these for a rigid-sphere model. Here, the potential is simply $V = \infty$ for $R \leq R_0$, $V = 0$ for $R > R_0$ (where R_0 is the sphere diameter). Then Eqn. 5.6 yields (independent of collision energy)

$$\theta = 2 \arccos(b/R_0) \qquad (b \le R_0) \tag{5.23}$$

and $\theta = 0$ for $b > R_0$, so that the differential cross section (Eqn. 5.22) becomes

$$\frac{d\sigma(\theta)}{d\omega} = \tfrac{1}{4} R_0^2 \tag{5.24}$$

i.e., isotropic and energy-independent. The total scattering cross section (resulting from integration, over all solid angles, of the differential cross section), namely

$$\sigma \equiv \int_{4\pi} d\omega \, \frac{d\sigma(\theta)}{d\omega} = 2\pi \int_0^{\pi} d\theta \sin\theta \, \frac{d\sigma(\theta)}{d\omega} \tag{5.25}$$

yields for the rigid-sphere model the expected result

$$\sigma = 2\pi \int_0^{\pi} d\pi \sin\theta \, \tfrac{1}{4} R_0^2 = \pi R_0^2 \tag{5.26}$$

(Using Eqn. 5.21b, $\sigma = \int d\sigma = \pi b_{max}^2 = \pi R_0^2$ in accordance with Eqn. 5.26.)

For the realistic potentials, such as those from Eqns. 5.1 and 5.2, however, problems arise immediately in the form of a divergence in the limit of $\theta \approx 0$ as the factor $(\sin\theta \, |d\theta/db|)^{-1}$ in Eqn. 5.22 approaches infinity and another at $\theta(b_r)$ where $d\theta/db = 0$. (The latter is responsible for the classical rainbow scattering phenomenon, as pointed out by E. A. Mason and O. Firsov.) Obviously, there is a flaw in the classical mechanical approach, and Eqn. 5.22 is therefore only a classical approximation (of very restricted applicability) to the proper quantum mechanical result for the differential scattering cross section. Nevertheless, classical mechanics does yield considerable insight into the collision process and has especial utility in dealing with inelastic and reactive collisions (whose quantal treatment is computationally difficult). Before leaving the classical description, let us note a few of its useful results in the field of elastic scattering.

For collisions of relatively large impact parameter, the scattering angle is governed mainly by the long-range interparticle interaction, i.e., by $V(R)$ at large R. For the typical asymptotic form $V(R) = -C_s R^{-s}$, Eqn. 5.6 becomes (in the limit of large b and small θ):

$$\theta \propto (C_s/E) b^{-s} \tag{5.27}$$

and so Eqn. 5.22 becomes

$$\frac{d\sigma(\theta)}{d\omega} = f(s)(C_s/E)^{2/s} \theta^{-(2+s)/s} \tag{5.28}$$

where $f(s)$ is a known constant. (Note the unfortunate divergence as $\theta \to 0$!)

For the usual case (closed-shell systems), $s = 6$ and Eqn. 5.28 becomes

$$\frac{d\sigma(\theta)}{d\omega} = f(6)\theta^{-7/3}(C_6/E)^{1/3} \tag{5.29}$$

Equation 5.29 suggests that a log-log plot of the c.m. differential scattering cross section (at fixed E) as a function of scattering angle would have a (limiting) linear slope of $-\frac{7}{3}$. This is indeed the observed behaviour (as shown by R. B. Bernstein and co-workers F. A. Morse, H. U. Hostettler, P. J. Groblicki, R. W. Bickes, K. G. Anlauf, and others) except that the very small angle divergence is absent; the value of $d\sigma/d\omega$ levels off to a definite, finite value as $\theta \to 0$ (first confirmed experimentally by R. K. Helbing and H. Pauly).

This is fortunate, since otherwise we would be faced with a very nonphysical result, namely, an infinite total scattering cross section (application of the integration of Eqn. 5.25 using Eqn. 5.29 for the low-angle differential scattering cross section). The quantal result is, of course, finite, as will be seen below. Equation 5.29 is useful in that it relates the long-range attractive C_6 potential constant to the magnitude of the low-angle scattering.

One final problem with the classical treatment is the unphysical divergence in the differential cross section at the rainbow angle. The rainbow phenomenon per se has a satisfactory quantum description, well confirmed by molecular beam scattering experiments, as mentioned later on. The classical treatment is of some use, nevertheless, in its predictions that the rainbow angle should vary inversely with the ratio of the collision energy to the potential well depth, i.e.,

$$\theta_r = c\epsilon/E \tag{5.30}$$

where the constant c depends only slightly upon the shape of the potential in the neighborhood of the well.

For large scattering angles dominated by the repulsive part of the potential, it is possible to deduce the latter from the observed differential cross section through the use of purely classical mechanics. For $\theta > \theta_r$ consider the 'partial' total cross section beyond θ, say $\sigma_>(\theta)$, defined as

$$\sigma_>(\theta) \equiv 2\pi \int_\theta^\pi d\theta \sin\theta \frac{d\sigma(\theta)}{d\omega} \tag{5.31}$$

Using Eqn. 5.22, for $d\sigma(\theta)/d\omega$,

$$\sigma_>(\theta) = 2\pi \int_0^\pi d\theta\, b \left|\frac{db}{d\theta}\right| = \pi \int_\theta^\pi d\theta \left|\frac{db^2}{d\theta}\right| = \pi[b(\theta)]^2 \tag{5.32}$$

Thus

$$b(\theta) = [\pi\sigma_>(\theta)]^{1/2} \tag{5.33}$$

The inversion procedure is as follows. Given $d\sigma(\theta)/d\omega$ for $\theta > \theta_r$, one obtains $\sigma_>(\theta)$ by integration according to Eqn. 5.31 and thus $b(\theta)$, from Eqn. 5.32, which is equivalent to $\theta(b)$. Knowing $\theta(b)$, one can determine $V(R)$ provided the collision energy is high enough such that no 'orbiting collisions' occur (typically, if $E > \epsilon$ this condition is met). The Firsov inversion procedure is as follows. Given $\theta(b)$, as determined above, one evaluates the integral $I(s)$, defined as

$$I(s) = \pi^{-1} \int_{b=s}^{\infty} db \, \frac{\theta(b)}{(b^2 - s^2)^{1/2}} \tag{5.34}$$

for a series of arbitrary values of the 'dummy parameter' s. It can be shown that

$$V(R) = E\{1 - \exp[-2I(s)]\} \tag{5.35}$$

where $R = s \exp[I(s)]$. $\tag{5.36}$

(Equations 5.35 and 5.36 together yield the following interpretation of s:

$$s = R[1 - V(R)/E]^{1/2} \tag{5.37}$$

but this is not required for the analysis.)

For each chosen field of s, one has a numerical value of $I(s)$ via Eqn. 5.34 and knows R from Eqn. 5.36. Equation 5.35 yields $V(R)$ for that value of R. The procedure is repeated for each s (and the corresponding R values), resulting in a unique set of points $V(R)$. This determines the repulsive part of the potential curve up to a potential energy equal to the collision energy E.

Enough of the virtues (and shortcomings) of the classical mechanical treatment. The exact quantum mechanical description of elastic scattering by a spherically symmetric potential is not only well understood but also readily executable by computer. What follows is a brief sketch of the 'practical' *modus operandi;* for details, see the monographs.

5.5. Quantal scattering theory

The quantum mechanical treatment of elastic scattering by a static potential uses the time-independent Schrödinger equation and the method of partial waves (as first applied to this field by N. F. Mott and H. S. W. Massey). Assuming a potential function $V(R)$, one represents the colliding system by an incident plane wave scattered by a centre of force to yield an outgoing spherical wave whose amplitude determines the differential scattering cross section.

The total wave function of the system is resolved into a sum of partial waves

$$\psi = \sum_{l=0}^{\infty} R_l(R) P_l(\cos\theta) \tag{5.38}$$

where l is the angular momentum quantum number, P_l the Legendre function, and $R_l(R)$ the radial wave function (BER66). The Schrödinger equation is set up and separated as usual, yielding an infinite set of radial equations of the form

$$\left[\frac{d^2}{dR^2} + k^2 - \frac{2\mu}{\hbar^2} V(R) - \frac{l(l+1)}{R^2} \right] G_l(R) = 0 \qquad (5.39)$$

where $k = (2\mu E / \hbar^2)^{1/2}$ is the incident wavenumber ($k = \mu v_r / \hbar = p/h = 2\pi/\lambda$) and $G_l(R) \equiv R \times R_l(R)$ is a modified radial function, a so-called partial wave, such that

$$G_l(0) = 0 \qquad (5.40)$$

In the absence of a potential ($V = 0$), $G_l^{(0)}$ is a spherical Bessel function whose asymptotic form is sinusoidal:

$$G_l^{(0)}(R) \simeq \sin(kR - \pi l/2) \qquad (5.41)$$

With the potential present, the solution of Eqn. 5.39 yields a wave function different from $G_l^{(0)}$, but also having an asymptotic sinusoidal form, phase shifted:

$$G_l(R) \simeq \sin(kR - \pi l/2 + \eta_l) \qquad (5.42)$$

The amplitude of the scattered wave at some angle θ from the direction of the incident plane wave is given by a sum over all the partial waves:

$$f(\theta) = \frac{1}{2ik} \sum_{l=0}^{\infty} (2l + 1)(e^{2i\eta_l} - 1) P_l(\cos\theta) \qquad (5.43)$$

The differential scattering cross section, or scattered intensity, is given by the mod-squared amplitude:

$$\frac{d\sigma(\theta)}{d\omega} \equiv I(\theta) = |f(\theta)|^2 = Re^2[f(\theta)] + Im^2[f(\theta)] \qquad (5.44)$$

This procedure was first implemented in connection with atom-atom scattering by R. B. Bernstein and further exploited by H. Pauly and co-workers.

In practice, the infinite sum in Eqn. 5.43 requires only a finite number of terms (e.g., 10^2–10^4) since the phase shifts tend to zero at sufficiently large l values. One can estimate the value of l, say l_{max}, beyond which simple analytical quadratures can replace summation. There is a semiclassical relation between the orbital angular momentum and the quantum number:

$$L = \mu v_r b = [l(l+1)]^{1/2} \hbar \qquad (5.45)$$

Thus

$$l \approx \mu v_r b / \hbar = kb \qquad (5.46a)$$

and

$$l_{\max} \approx k b_{\max} \qquad (5.46b)$$

From Eqn. 5.21b, the total scattering cross section is

$$\sigma = \int d\sigma = \pi b_{\max}^2 \qquad (5.47)$$

and so, from Eqn. 5.46b we estimate

$$l_{\max} \approx k(\sigma/\pi)^{1/2} \qquad (5.48)$$

In the simplest case of the rigid-sphere potential, from Eqn. 5.26 we find

$$l_{\max} \approx kR_0 \qquad (5.49)$$

However, for realistic potentials with long-range attractive tails, for which the cross section σ is much larger than πR_e^2, $l_{\max}$ is typically two or four times larger than kR_e.

There are well-known procedures to compute the set of phase shifts from the radial Eqn. 5.39, as described in the monographs. Suffice it to say here that for realistic intermolecular potentials, such as those of Eqns. 5.1 and 5.2, the dependence of η upon l and E is well understood, and semiclassical methods for their approximate calculation (and for their descriptive analysis) are well developed. (The ground-breaking work of K. W. Ford and J. A. Wheeler was followed by the practical implementation methodology of R. B. Bernstein and others.)

The quantum mechanically calculated differential scattering cross sections turn out to be finite in the limit of $\theta = 0$, show no divergence at θ_r, and yield upon integration finite and well-behaved total cross sections.

Figure 5.11 shows typical quantal-calculated differential cross sections for a Lennard–Jones (12,6) potential of given well depth ϵ and range parameter R_e. Shown is a comparison of the oscillatory quantal angular distributions (the fine oscillations due to de Broglie 'diffraction,' the coarse oscillations corresponding to rainbow interferences, etc.) with the smooth but ill-behaved classical differential cross sections. The three quantal calculations shown correspond to a constant value of the reduced collision energy $K = E/\epsilon$, but different values of the 'reduced' wave number $A \equiv kR_e \propto R_e/\lambda$. The larger the value of A, the smaller the de Broglie wavelength of the collision (relative to molecular dimensions) and thus the more 'classical' the behaviour of the system. Over a wide range of angles, the classical calculation of $d\sigma/d\omega$ shows the general trend of the quantal result, averaging out all the interference effects, however.

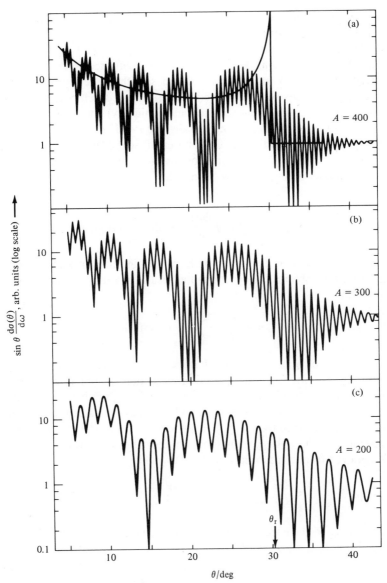

FIG. 5.11. Quantal differential elastic cross sections for a Lennard–Jones (12,6) potential corresponding to collisions at the same reduced energy ($E/\epsilon = 4$) but at different values of the reduced wavenumber $A \equiv kr_e \propto r_e/\lambda$. Large A implies 'more classical.' The classical differential cross section, however, is independent of A and is shown as the smooth curve in (a). Adapted from H. Pauly, PAU79.

Modern molecular beam experiments using well-collimated crossed beams with narrow velocity distributions and sensitive detectors with good angular resolution (such as those from the laboratories of Y. T. Lee, H. Pauly, G. Scoles, A. Kuppermann, and others) have confirmed such theoretically expected behaviour. Various empirical and semitheoretical procedures have been successfully developed to 'invert' experimental differential cross section curves at several collision energies to yield (uniquely) the interparticle interaction potentials $V(R)$. For further details, see PAU79.

Returning to the equation for the differential scattering cross section (Eqn. 5.44), we can obtain a simple expression for the total cross section. Integration is accomplished analytically by using the orthonormal properties of the Legendre functions. From Eqn. 5.25,

$$\sigma = 2\pi \int_0^\pi d\theta \sin\theta \, \frac{d\sigma(\theta)}{d\omega} = 2\pi \int_0^1 dx |f(x)|^2 \qquad (5.50)$$

where $x \equiv \cos\theta$ and $f(x)$ is given by Eqn. 5.43 in terms of the sum of $P_l(x)$ terms. One then obtains the exact quantal result for the total elastic scattering cross section:

$$\sigma = \frac{4\pi}{k^2} \sum_{l=0}^{\infty} (2l + 1)\sin^2\eta_l \qquad (5.51)$$

The behaviour of η as a function of l and E thus governs both the angular distribution of scattering and the total cross section. To obtain an estimate of σ, we note that for realistic systems at moderately high reduced wavenumbers ($A \gg 1$), $\eta_0 \approx -A$, $\Delta\eta_l/\Delta l \approx \pi/2$, and so $\sin^2\eta_l$ oscillates rapidly between 0 and 1 for l changing from $l = 0$ to some large value, the 'cutoff,' say L. In the so-called random-phase approximation, the expectation value of $\sin^2\eta_l$ (i.e., ½) is removed from the sum in Eqn. 5.51 to yield

$$\sigma_{<L} \equiv \sigma(l \le L) \approx \frac{2\pi}{k^2} \sum_{l=0}^{L} (2l + 1) \approx \frac{4\pi}{k^2} \sum_{l=0}^{L} l = \frac{2\pi L^2}{k^2} = 2\pi b_L^2 \qquad (5.52)$$

Here $b_L = L/k$, the impact parameter corresponding to the cutoff $l = L$. For the rigid-sphere case, $b_L \approx R_0$, and so

$$\sigma_{<L} \approx 2\pi R_0^2 \qquad (!) \qquad (5.53)$$

The nonintuitive factor of 2 is understood in terms of the forward 'diffraction peak' in the differential cross section (whose integrated contribution to the total cross section is πR_0^2), termed the 'shadow scattering' effect, as first pointed out by H. S. W. Massey.

Returning to the realistic case of the inverse 6th-power long-range potential, one can well approximate the higher-order phase shifts by the Born approximation, with the result:

$$\eta_l = \frac{3\pi}{32} \left(\frac{2\mu}{\hbar^2} C_6 \right) k^4 l^{-5} \equiv a \, l^{-5} \qquad (5.54)$$

For sufficiently large l, $\eta_l \ll \pi/2$ and $\sin\eta_l \approx \eta_l$. Substituting Eqn. 5.54 into Eqn. 5.51 and using Eqn. 5.52, we obtain for the total cross section ($\sigma = \sigma_{<L} + \sigma_{>L}$)

$$\sigma = \frac{4\pi}{k^2} \left[\frac{L^2}{2} + \int_L^\infty d\,l (2\,l + 1)\sin^2\eta_l \right]$$

$$\approx \frac{2\pi}{k^2} \left(L^2 + 4 \int_L^\infty d l \, l\eta_l^2 \right) = \frac{2\pi}{k^2} \left(L^2 + \frac{a^2}{2} L^{-8} \right) \qquad (5.55)$$

$$= \frac{2\pi L^2}{k^2} (1 + \eta_L^2/2) = \frac{2\pi}{k^2} \left(\frac{a}{\eta_L} \right)^{2/5} (1 + \eta_L^2/2)$$

The Massey–Mohr (MM) approximation to the cross section assumes $\eta_L = \frac{1}{2}$ so that

$$\sigma_{MM} = \left(\frac{2\pi}{k^2} \right) k^{8/5} \left(\frac{3\pi}{16} \right)^{2/5} \left(\frac{2\mu C_6}{\hbar^2} \right)^{2/5} \left(\frac{9}{8} \right) \qquad (5.56)$$

$$= \frac{9\pi}{4} \left(\frac{3\pi}{8} \right)^{2/5} \left(\frac{C_6}{\hbar v_r} \right)^{2/5}$$

Thus the cross section should decrease with increasing relative velocity v_r (with a $v_r^{-2/5}$ dependence) and its magnitude should be governed by C_6. For details, see BER66.

Figure 5.12 shows a full quantal calculation of the velocity dependence of the total elastic scattering cross section, for the LJ (12,6) potential, in reduced units, i.e., a log-log plot of $\sigma(v_r)$. At low relative velocities, the behaviour is largely determined by the R^{-6} long-range attraction, accounting for the slope of $-\frac{2}{5}$; at high collision energies the repulsive, short-range R^{-12} potential dominates and the slope is $-\frac{2}{11}$.

The oscillatory deviations from the monotonic MM behaviour at low energies arise from the so-called glory effect, having to do with interference between the forward ($\theta \approx 0$) scattered rays from large b and those from b near the glory impact parameter b_g (Fig. 5.10) for which $\theta \approx 0$ also. The velocities of the glory extrema can be used to estimate the potential well characteristic product ϵR_e, pointed out by R. B. Bernstein and T. J. O'Brien, and demonstrated experimentally by E. W. Rothe and P. K. Rol, as discussed in the monographs. Experimental studies of the glory effect for many atom-atom

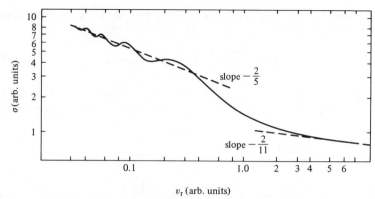

v_r (arb. units)

FIG. 5.12. Log-log plot of total elastic cross section versus relative speed v_r for a LJ (12,6) potential, showing glory extrema at low energies superimposed on the monotonic $v^{-2/5}$ behaviour for an R^{-6} long-range attraction. At high velocities, the asymptotic slope of $v^{-2/11}$ corresponds to the R^{-12} short-range repulsion. Adapted from R. B. Bernstein, *J. Chem. Phys.* **34**, 361 (1961) and H. Pauly, PAU75.

collision systems have confirmed the theory (fully developed by R. Düren and H. Pauly) in every detail. However, for molecular systems, the amplitude of the glories are often 'quenched' because of the anisotropic character of the potential, as discussed by R. E. Olson, R. B. Bernstein, R. J. Cross, J. Reuss, and others. Suffice it to say here that the influence of noncentral terms in the potential is important, as they not only tend to 'smear out' interference effects (e.g., in both the differential cross section and glories) but also induce rotational inelasticity in the collisions. This field is of great interest in its own right, but we shall have to limit our discussion here.

Another interesting quantal interference effect arises when the two collision partners are indistinguishable, e.g., identical isotopes. Figure 5.13 shows experimental results (and quantal calculations from a single best-fit potential) for the total elastic scattering cross sections (versus v_r) of the three isotopic systems ^{3}He-^{3}He, ^{3}He-^{4}He, and ^{4}He-^{4}He. Both qualitative and quantitative features of the experimental results are accounted for by the quantal scattering theory, the quantitative aspects requiring a good potential curve $V(R)$ at least up to energies V corresponding to the highest collision energy E of the experiments.

Modern inversion methods, based on the work of O. Firsov and of U. Buck and H. Pauly, have made it possible to deduce many interatomic (and even intermolecular) potentials from elastic molecular beam scattering experiments. These potentials have been tested by calculation of bulk properties, such as transport and virial coefficients (the results are indeed found to be consistent) and bound states for the van der Waals molecules (whose eigenvalues have been measured spectroscopically). Direct inversion of temperature-dependent transport and virial coefficients using procedures of J. S. Rowlinson, E. B. Smith, G. C. Maitland, and others has yielded potentials in good accord with these. On the whole, it is found that the molecular beam derived potentials are entirely satisfactory and second in accuracy only to spectroscopically determined potentials (usually limited to the region of the minima).

5.6. Inelastic scattering

Turning now to inelastic collisions, we shall briefly touch upon the measurement of rotational and vibrational energy transfer induced by molecular impact, as detected with molecular beam methods.

The most direct measurements of collisionally induced rotational transition cross sections make use of state-selected molecular beams and use state analysis of the molecules scattered at various angles. This technique was pioneered by J. P. Toennies and co-workers.

A somewhat more general technique for determining the inelasticity in a

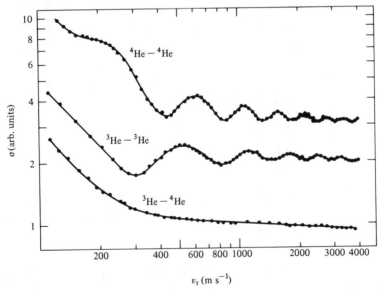

FIG. 5.13. Experimental (points) versus calculated (solid curves) velocity dependence of total elastic cross sections for isotopic He collisions. Adapted from H. Pauly, in BER79, p. 111, from R. Feltgen, H. Pauly, F. Torello, and H. Vehmeyer, *Phys. Rev. Lett.* **30**, 820 (1973).

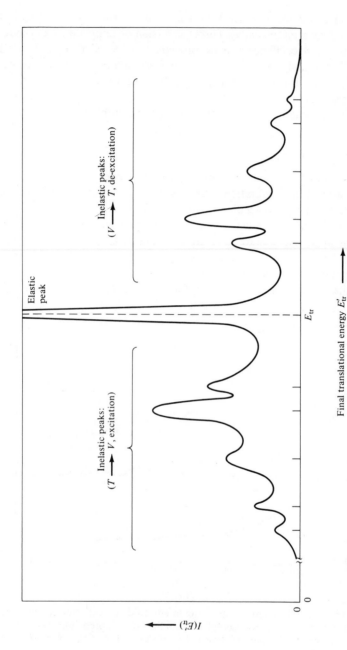

Fig. 5.14. Hypothetical translational energy spectrum for vibrationally inelastic collision, showing strong elastic scattering peak surrounded by Stokes and anti-Stokes spectral features. See text.

collision is the velocity-change method, first applied to molecular beams by R. B. Bernstein and co-workers A. R. Blythe and A. E. Grosser, and beautifully expolited, for slow ion beams, by J. P. Toennies et al. Figure 5.14 is a highly schematic drawing of a hypothetical translational energy spectrum for a vibrationally inelastic molecular collision, showing the strong elastic peak at E'_{tr} = E_{tr}, the various vibrational excitation ($T \to V$) peaks for which $E_{tr} - E'_{tr}$ = $n\hbar\omega_i$, and the analogous ($V \to T$) de-excitation peaks. We see immediately a strong analogy; here we have a translational Raman spectrum, with an intense Rayleigh line, with Stokes and anti-Stokes lines shifted by characteristic energy displacements from the Rayleigh frequency.

In practice, one measures the velocity distribution of one or the other of the collision partners at various scattering angles. In the favourable case of nearly monochromatic, collimated, crossed beams, the time-of-flight velocity analysis technique allows an accurate determination of the differential (c.m.) cross section for given ΔE ($\equiv E'_{tr} - E_{tr}$).

Figure 5.15 shows a velocity vector diagram intended to represent a rotationally inelastic collision of a heavy atom with a D_2 molecule.

The 'state-to-state' differential (c.m.) cross section (obtained by such molecular beam experiments) is one of the most detailed sources of information bearing upon the anisotropic intermolecular potential surface. From a practical

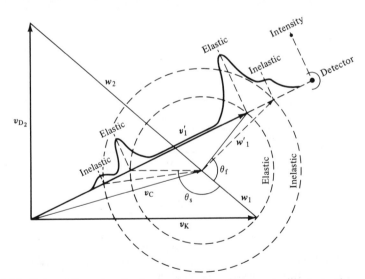

FIG. 5.15. Schematic and sylized velocity vector diagram to illustrate the velocity analysis technique for the study of rotationally inelastic scattering. The process is de-excitation of D_2 ($j = 2 \to j' = 0$) by collision with K. The solid curve is the expected velocity distribution of scattered K at the given angle. Adapted from A. R. Blythe, A. E. Grosser, and R. B. Bernstein, *J. Chem. Phys.* **41**, 1917 (1964); details therein.

viewpoint, however, the inversion of such cross section data to the potential is still difficult, and the most common procedure is to assume reasonable functional forms (such as Eqn. 5.4) for the potential, calculate $d\sigma_{j \to j'}(\theta)/d\omega$ (via semiclassical or, preferably, quantal methods), compare with experiment, and iterate to a best fit of the measurements.

An example of a well-studied system is shown in Fig. 5.16. Plotted are time-of-flight distributions for HD scattered by He at various laboratory angles, in which individual final rotational states are resolved. Figure 5.17 shows resolved differential (c.m.) cross sections for the HD + D_2 system, for transitions in HD from $j = 0$ to 1 (and for the pure elastic case, $j = 0 \to 0$). The sum of these cross sections accounts for essentially all the (total) scattering at the low collision energy of the experiments. Such informative experiments are the result of such significant technical advances as those of W. R. Gentry and C. F. Giese and of U. Buck and H. Pauly and others.

5.7. State-to-state cross sections

A very recent development in the field of inelastic, state-to-state scattering has been possible through the use of advanced laser technology. By exciting a diatomic molecule with one laser and then probing the scattered molecules with

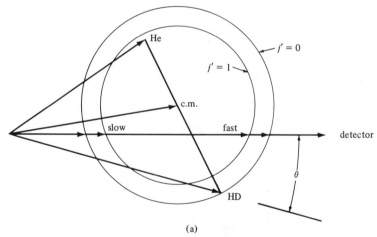

(a)

FIG. 5.16a. Velocity vector diagram appropriate for the inelastic scattering of the He-HD system as studied in a special crossed beam machine.

FIG. 5.16b. Fully resolved TOF spectra of HD scattered by the He at specified angles as defined in Fig. 5.16a. Path length 0.48 m. Adapted from W. R. Gentry and C. F. Giese, *J. Chem. Phys.* **67**, 5389 (1977); details therein.

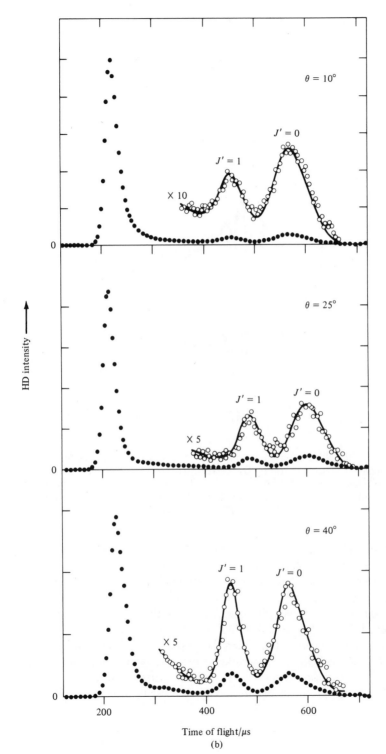

$\theta = 10°$

$J' = 1$

$J' = 0$

× 10

$\theta = 25°$

$J' = 1$

$J' = 0$

× 5

$\theta = 40°$

$J' = 1$

$J' = 0$

× 5

HD intensity ⟶

0

0

0

200   400   600

Time of flight/μs

(b)

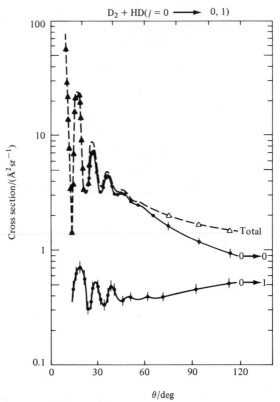

FIG. 5.17. Resolved differential cross sections, transformed to the c.m. system, for the rotational excitation of HD by D_2. *Lower curve: $j = 0 \to 1$; upper (solid) curve: $j = 0 \to 0$ (elastic)*; triangles are sum of $0 \to 0$ and $0 \to 1$ cross sections; dashed curve represents calculated total differential cross sections. Adapted from U. Buck, F. Huisken, and J. Schleusener, *J. Chem. Phys.* **68**, 5654 (1978).

another, one can measure state-to-state total inelastic cross sections. Very narrow laser linewidths are available via single-mode operation of certain tunable dye laser systems. This has made possible the observation of Doppler shifts (and determination of Doppler widths) in the laser-induced fluorescence, so that the laboratory velocity distribution of the scattered molecules can be deduced. Figure 5.18 is a sketch of such an apparatus.

In favourable cases, it has been shown (by J. L. Kinsey and D. E. Pritchard and by K. Bergmann and co-workers) that one can extract from such data the *angular* distribution of rotational transition cross sections.

Direct measurements of *vibrational* inelasticity in neutral molecular beam experiments have been difficult (partly because of the small cross sections for collisions at low, 'epithermal' energies). However, analogous experiments with charged particles are much easier, and extensive studies of inelastic ion-molecule scattering have been carried out by J. P. Toennies and co-workers. Figure 5.19 shows early results of vibrational transition probabilities for the $H_2 + Li^+$ system, at two collision energies, plotted for various excitations of the H_2 ($v = 0 \rightarrow 0,1,2,3,4,$) as a function of the c.m. scattering angle. With the availability of detailed dynamic data, it is possible to test various *ab initio* calculated potential energy surfaces for the $H_2 + Li^+$ system, $V(R,\gamma)$, e.g., Fig. 5.4 (which has since been supplanted by a more accurate surface, consistent with the inelastic scattering results). More recently, state-to-state angle-dependent inelastic cross sections have been measured for simple polyatomic molecules by J. P. Toennies and co-workers.

In concluding this chapter, it would be well to recall that molecular beam scattering is only one method for investigating potential energy surfaces. Very accurate experimental determinations of anistropic potential wells have been carried out using modern spectroscopic techniques (e.g., the experiments on

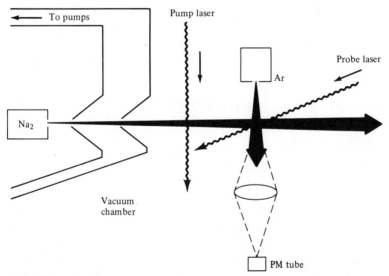

FIG. 5.18. Schematic diagram of laser molecular beam apparatus for measurement of state-to-state differential cross sections of vibrotationally inelastic collisions of Na_2 with Ar. The 'pump and probe' technique is used. Adapted from J. A. Serri, C. H. Becker, M. B. Elbel, J. L. Kinsey, W. P. Moskowitz, and D. E. Pritchard, *J. Chem. Phys.* **74**, 5116 (1981); details therein. See also J. A. Serri, A. Morales, W. Moskowitz, D. E. Pritchard, C. H. Becker, and J. L. Kinsey, *J. Chem. Phys.* **72**, 6304 (1980) and K. Bergmann, U. Hefter, and J. Witt, *J. Chem. Phys.* **72**, 4777 (1980).

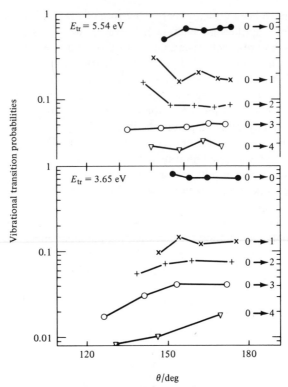

F<small>IG</small>. 5.19. Vibrational transition probabilities as a function of c.m. scattering angle for $H_2(v = 0 \rightarrow v' = 0,1,2,3,4)$ excited by collision with Li^+ at the indicated values of the relative translational energy. Adapted from H. E. Van den Bergh, M. Faubel, and J. P. Toennies, VAN73.

isotopic hydrogen molecule–rare gas systems by H. L. Welsh, as analyzed by R. J. LeRoy and J. Van Kranendonk, and others). The molecular beam electric resonance method for the study of polar diatomic–rare gas van der Waals molecules (pioneered by W. Klemperer and L. Wharton) has yielded accurate structural data and potential surface characteristics (cf. Fig. 5.5). Finally, one must expect that *ab initio* quantum mechanically computed surfaces will soon be available (at least for few-electron systems) with an accuracy sufficient to be *predictive* of experiment. Thereafter, inversion schemes to deduce $V(R,\gamma)$ from scattering and spectroscopic data will be useful only for more complicated systems (those for which computations become more costly than experiments!). Clearly, molecular beam and laser techniques will continue to serve as a principal methodology for the study of intermolecular forces.

Review literature

MOT33, FOR59, ROS62, BER64, BER64a, GOL64, MCD64x, PAU65, TAK65,
AMD66, BER66, GRE66, ROS66, BER67, BER67a, HIR67, HAR68, PAU68,
SCH69, LES70, ROS70, SCH70, MAS71, KIN72, LEE72, SMI72, FLU73,
GOR73, PAU73, VAN73, WON73, FAR74, SEC74, TOE74, TOE74a, BUC75,
LAW75, MIC75, REU75, KUN76, LES76, MIL76, TOE76, BER79, PAU79,
STO79, HEN81, MAI81, MIC81, TRU81

6 Reactive scattering on adiabatic potential surfaces

6.1. Potential energy hypersurfaces appropriate for reactive scattering

We have seen how the potential energy surface governs the elastic and inelastic scattering behaviour of atoms and molecules, and how information on the potential can be extracted from measurements of various scattering cross sections. At least for three-atom systems, such data (supplemented by spectroscopic results) can be fairly well 'inverted' to yield the potential, albeit over limited ranges of the variables, as discussed in Chapter 5. Of course, one would prefer a quantum mechanically computed potential surface with global accuracy, but the empirically derived potentials, when available, are often competitive in a practical sense.

When we consider reactive collisions, however, the problem of inversion of scattering data to yield the potential energy hypersurface is even more difficult, and so one has an even greater need for *ab initio* potentials. Although it is naive to believe that reactive scattering data can ever really define a potential energy hypersurface, it has hitherto been customary to attempt what might be termed a 'partial inversion,' deducing key topological aspects of the potential from salient features of the scattering. This was the case because all the observable molecular dynamics stem from the potential surface. Some of the major dynamic features, such as population inversion, direct versus complex mode behaviour, role of translational versus internal energy, can be traced to specific 'responsible' features of the potential surface.

Since the pioneering work of J. O. Hirschfelder and H. Eyring more than a half-century ago, classical mechanical trajectories on assumed potential hypersurfaces have been helpful in explaining the gross features of reaction dynamics for a large number of elementary reactions. (Recall the discussion of Figs. 2.8 and 2.9 and the subsequent material in Chapter 2.) For example, direct-mode reactions usually have potential barriers in the reaction coordinate, whereas for complex-mode reactions a potential well in the surface 'traps' the combining reagents for a time sufficient to lose the memory of some of the details of the initial encounter. Before going into more detail on such 'micromechanisms' of chemical reaction, it is worthwhile to go back to very primitive considerations and discuss the following question: what can we learn, *qualitatively,* from a measured angular distribution of reactive scattering?

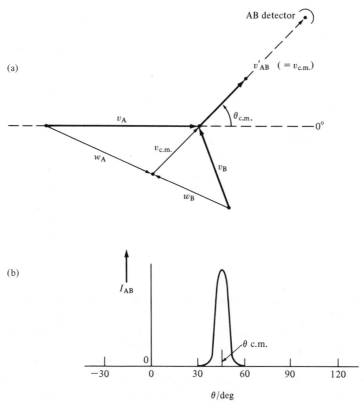

FIG. 6.1. (a) Velocity vector diagram for the formation of a very long-lived complex or adduct consisting of the two colliding molecules A and B; (b) angular distribution of the scattered flux, a sharp peak located at the centroid angle. See text.

6.2. Extremes of behaviour for reactive scattering: complex versus direct mode

The simplest possibility would be the case where the reagent molecules, say A and B, would simply stick together to form a long-lived complex, or adduct, A $\cdots$ B, as portrayed in the velocity vector diagram of Fig. 6.1a. If there are enough vibrational degrees of freedom in the adduct to 'accept' the exoergicity of the combination reaction, the lifetime of the complex can exceed the microsecond to millisecond transit time required to reach the detector, and one would observe the adduct directly at the centroid angle, as shown in Fig. 6.1b. (The small angular width of the peak is due to an unavoidable spread in relative velocity and necessarily imperfect beam collimation.) This is rarely the case, at least for systems studied thus far.

However, a more common situation is one in which the AB adduct rearranges in a short time τ and falls apart to give C and D, or A and B. This is the so-called long-lived complex mechanism,

$$A + B \rightleftarrows [AB] \rightarrow C + D \qquad (6.1)$$

first observed by D. R. Herscnbach and co-workers. It is of interest to measure the branching ratio for AB decay into C + D versus A + B.

For simplicity, we shall work in the centre-of-mass system: let us now consider the recoil of the products with respect to the centre-of-mass (c.m.). Figure 6.2a shows the incident antiparallel relative velocity vectors w_A and w_B and possible cones of recoiling product velocity vectors w_C and w_D (relative to the c.m.). For the case in which the complex lives for a sufficient time (e.g., a number of picoseconds) to rotate several revolutions, the decay products are distributed with the same probability in the 'forward' as in the 'backward' direction (with respect to w_A, for example). Thus the angular distribution of each of the products C and D is symmetric about 90° with respect to the direction of the incident relative velocity vector v_r ($\equiv w_A - w_B$), and of course that of C is the same for D.

Figure 6.2b is intended as an aid in visualization of a plausible product velocity vector distribution in the c.m. system for the case of the decay of a long-lived complex. Figure 6.2c shows a typical product angular distribution (in the c.m. system), where the ordinate is the differential solid angle cross section, proportional to the flux of product at the given scattering angle, per steradian. The total yield in the forward and backward hemispheres is equal: the products have 'forgotten' the original direction of v_r. Note, however, that they have *not* forgotten the total angular momentum of the colliding reagents A and B. The implications of angular momentum conservation with regard to product angular distribution are well known but beyond the scope of this discussion. (For a further discussion at an elementary level, see MRD.) The first experimental demonstration of this long-lived 'complex-mode' mechanism for an elementary reaction and its theoretical basis was by D. R. Herschbach, W. B. Miller, and S. A. Safron.

An experimental result, for the reaction $O + Br_2 \rightarrow OBr + Br$, is shown in Fig. 6.3. Even without any analysis, the forward-backward symmetry of the OBr confirms that the mechanism of the reaction involves the formation of a relatively long-lived complex.

Next we shall consider the case of a so-called direct-mode reaction, one that occurs on a much shorter (e.g., subpicosecond) time scale, one that is 'over' before the combined system AB has time to rotate. This 'direct mode' of reaction is exemplified by the $H + H_2$ atom transfer reaction, for which early trajectory simulations predicted interaction times (i.e., lifetimes of the H_3 'complex') of less than about 10^{-13} s. Direct-mode reactions display strongly

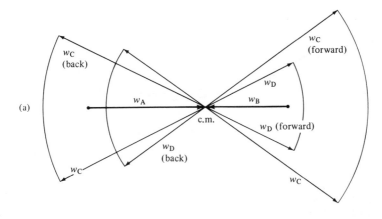

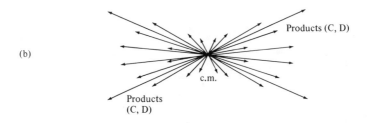

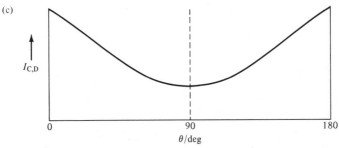

Fig. 6.2. (a) Centre-of-mass velocity vector diagram showing the collision of molecules A and B followed by the symmetric decay of the complex to yield molecules C and D (of unequal masses). Two possible angles (cones) of recoil are shown, as labelled; (b) symmetric (forward-backward) emission pattern of products in c.m. system; (c) centre-of-mass angular distribution function for the products. See text.

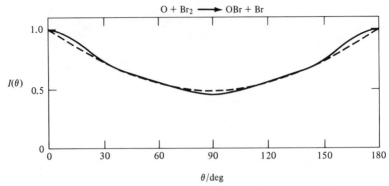

FIG. 6.3. *Solid curve:* c.m. angular distribution for OBr product from the crossed beam reaction of O + Br₂. *Dashed curve:* theoretical fit based on statistical complex model. Adapted from D. D. Parrish and D. R. Herschbach, *J. Amer. Chem. Soc.* **95**, 6133 (1973).

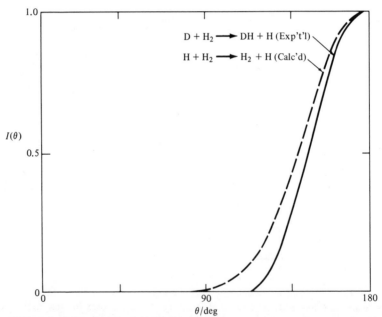

FIG. 6.4. *Solid curve:* experimentally derived c.m. angular distribution of reactive scattering of HD product from the D + H₂ reaction. Adapted from J. Geddes, H. F. Krause, and W. L. Fite, *J. Chem. Phys.* **56**, 3298 (1972). *Dashed curve: ab initio* calculated angular distribution for the H + H₂ exchange reaction. Adapted from A. Kuppermann and G. C. Schatz, *J. Chem. Phys.* **62**, 2502 (1975); **65**, 4668 (1976).

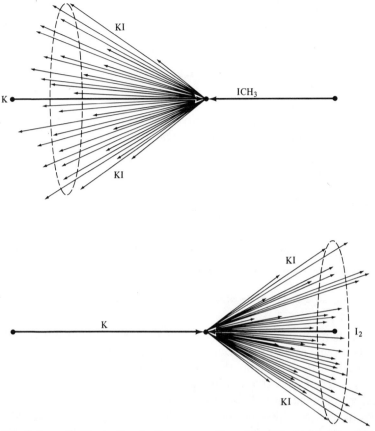

FIG. 6.5. Schematic illustration in the c.m. system of backscattering of product of rebound reaction (e.g., KI from K + ICH$_3$) and the forward-scattering of product from stripping reaction (e.g., KI from K + I$_2$). The cones shown contain most (but not all) of the scattered flux of the diatomic product. See MRD, Section 3.3, for details.

anisotropic c.m. angular distributions of products, with the shape of the angular distribution a rather direct manifestation of key dynamic factors, e.g., the preferred internuclear configuration of the transition state. Figure 6.4 shows the angular distribution for the H$_2$ molecular product of the H + H$_2$ reaction, which is sharply peaked at 180° (backscattered relative to the incident atom direction).

Figure 6.5 shows schematically the salient features characterising two extreme varieties of direct-mode reactions, the 'rebound' and the 'stripping' types. We shall consider the atom-molecule exchange reactions of the type

$$M + XR \rightarrow MX + R \qquad (6.2)$$

where M is a metal atom, X a halogen atom, and R an alkyl radical (or another halogen, or H), as first studied by S. Datz and co-workers, by D. H. Herschbach et al., by R. B. Bernstein and co-workers, and others.

Later we shall look at experimental data for several reaction systems represented by Eqn. 6.2, but it is clear that, at the very least, the shape of the c.m. angular distribution of scattered reaction products tells us whether the reaction occurs *directly* (in less that a ps or so) or via a 'sticky collision.' Such information was not available before the advent of the molecular beam scattering technique and its application to chemical reaction kinetics, pioneered by E. H. Taylor and S. Datz, D. R. Herschbach, and others.

6.3. Cross sections and reaction probability

If one integrates the product yield (more properly, the differential reaction cross section) over all solid angles, the total, or 'integral,' reaction cross section σ_R at the given collision energy can be obtained. This is a measure of the 'overall' reaction probability, via the expression

$$\sigma_R \equiv \int_0^\pi d\theta \, 2\pi \sin\theta \, \frac{d\sigma_R(\theta)}{d\omega} = \int_0^\infty db \, 2\pi b P_R(b) \qquad (6.3)$$

where the first equality is simply the defining relation between differential and total cross sections and the second the result of a classical interpretation in which $P_R(b)$ is the probability that a collision with impact parameter b results in reaction. The term $P_R(b)$ is also called the 'opacity function.' Of course, P_R represents an orientation-averaged probability, since in most experimental situations the relative orientation of the colliding molecules is not controlled. Figure 6.6 shows some plausible $P(b)$ functions. By its very nature, as an integral (Eqn. 6.3), σ_R cannot tell as much about the reaction as the opacity function $P(b)$ since an infinite variety of $P(b)$ curves can integrate up to the same σ_R. (Early work on the opacity function based on an 'optical model' treatment of the *non*reactive scattering was carried out by J. Ross and E. F. Greene.)

At a certain (low) level of approximation, assuming a unique relation between scattering angle and impact parameter, i.e., $b(\theta)$ functionality at the given E_{tr}, the integrands in Eqn. 6.3 can be equated such that

$$b(\theta) P_R[b(\theta)] = \sin\theta \, \frac{d\sigma_R(\theta)}{d\omega} \qquad (6.4)$$

and $P_R(b)$ evaluated from $d\sigma_R(\theta)/d\omega$. It is more common to attempt to fit observed differential and then integral cross sections via theoretically or semiempirically generated opacity functions than to attempt inversion of cross section data.

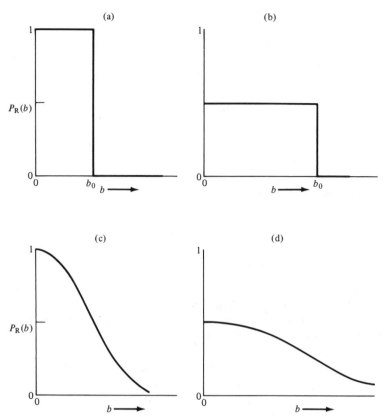

FIG. 6.6. Comparison of several stylized opacity functions, i.e., reaction probability P_R versus impact parameter b, at a given collision energy. Graphs (a) and (b) represent step functions, (c) and (d), respectively, more realistic opacity functions; (d) shows the common fact that the maximum orientation-averaged reaction probability is significantly less than unity. See MRD, Sections 2.5 and 3.3.

6.4. Translational energy dependence

Clearly, the reaction probability depends strongly upon the collision energy E_{tr}. (The opacity function P_R depends as well upon the internal quantum states of the reagent molecules, but the internal energies of beam molecules are more difficult to control.) The translational energy dependence of the reaction probability is a consequence of the shape of the potential energy hypersurface, as revealed by many classical trajectory computations of model systems. The integrated reaction cross section is therefore governed by the same factors, and its dependence upon E_{tr} can yield qualitative insight into the 'micromechanism' of the reaction. The first measurements of the translational energy dependence of a crossed molecular beam reaction cross section were those of M. E. Gersh and R. B. Bernstein for the $CH_3I + K$ reaction.

Figure 6.7 shows some theoretical cross section functions $\sigma_R(E_{tr})$. For endoergic reactions or those with an activation barrier, a 'threshold' for reaction exists, i.e., a minimum collision energy E_{th} is needed to achieve a measureable extent of reaction. The determination of threshold energies is in itself a valuable contribution to the characterisation of the reaction. Noting that

$$E_{th} = E_b + \Delta E_0 \tag{6.5}$$

where E_b (≥ 0) is the activation barrier and ΔE_0 the endoergicity of the reaction, any measured E_{th} represents an upper bound on the endoergicity. For an atom-molecule reaction of the type

$$A + BC \rightarrow AB + C \tag{6.6}$$

the dissociation energy of the product molecule is given by

$$D_0(AB) = D_0(BC) - \Delta E_0 = D_0(BC) + E_b - E_{th} \tag{6.7}$$

Thus the quantity $D_0(BC) - E_{th}$ represents a lower bound on the dissociation energy of the AB product (often an unknown quantity). Such translational threshold experiments have been carried out in the laboratories of Y. T. Lee, R. B. Bernstein, P. R. Brooks, R. J. Cross, M. Menzinger, A. G. Ureña, and others.

Referring back to the cross section function $\sigma_R(E_{tr})$, e.g., Fig. 6.7, one notes that for the 'line-of-centres' functionality

$$\sigma_R(E_{tr}) = \sigma^0(1 - E_{th}/E_{tr}) ; \quad (E_{tr} \geq E_{th}) \tag{6.8}$$

(where σ^0 is the maximum cross section), one obtains (under certain well-defined assumptions) the 'collision-theory Arrhenius equation' for the rate coefficient (cf. Eqn. 2.4):

$$k(T) = \langle v_r \rangle_T \, \sigma^0 \exp(-E_{th}/kT) \tag{6.9}$$

For details, see Chapter 2 and the monographs; suffice it to say that experimental cross section functions $\sigma_R(E_{tr})$ yield considerably more insight into the micromechanism of the reaction than Arrhenius parameters obtained from bulk rate data (recall Chapter 2).

For exoergic reactions (at least for those with essentially no activation barrier), the cross section functions often show a decrease with increasing collision energy, also shown in Fig. 6.7. At high energies, the effect of increasing the relative velocity is mainly to reduce the interaction time of the reagents and lower the overall reaction probability. This is found to be the case for many complex-mode reactions and for a large class of ion-molecule reactions. For the

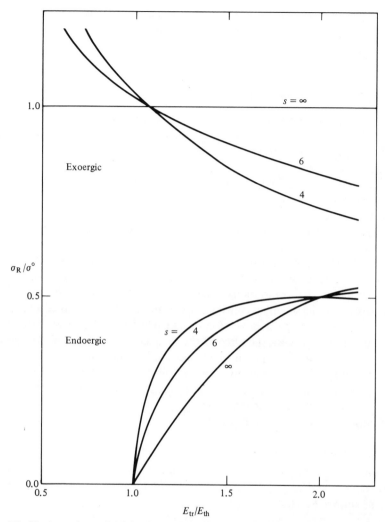

FIG. 6.7. Various theoretical functionalities for the translational energy dependence of the reaction cross section. The endoergic reactions have thresholds E_{th}; the reverse, exoergic reactions have energy dependencies of the form $\alpha_R \propto E_{tr}^{-2/s}$, where s is the power of the long-range attractive part of the interaction potential. The case of $s = \infty$ corresponds to the line-of-centres or Arrhenius-like cross section function. Adapted from R. D. Levine and R. B. Bernstein, *J. Chem. Phys.* **56**, 2281 (1972). See MRD, Section 2.5.

latter, it was long ago shown by G. Gioumousis and D. P. Stevenson that σ_R $\propto E_{tr}^{-1/2}$, and so the rate coefficient becomes temperature-independent (see MRD for details).

6.5. Kinematics of reactive collisions

Before going on with examples of reactive scattering from the molecular beam literature, it is well to review a few of the kinematic aspects of reactive collisions, i.e., the scattering in the laboratory frame and its relation to the c.m. scattering cross sections.

Let us now have a look at the qualitative features of product flux-velocity-angle contour diagrams. Figure 6.8 shows the velocity vector diagram for a hypothetical crossed beam experiment on an exoergic direct-mode reaction in which the reagent velocities are v_1 and v_2 and the detected product is species 3. In Fig. 6.8a it (3) is ejected predominantly in the 'forward' direction (with species 1 defining direction). Furthermore, the reaction is assumed to release only a small fraction of the total available energy into relative translation of the products. The translational energy released is some value E_{tr}', which exceeds E_{tr} by the so-called collisional exoergicity Q (here denoted Q_a). Shown is an arc representing a projection on the plane of the relevant part of the sphere of radius w_3 corresponding to Q_a. Analogously, in Fig. 6.8b, the product species 3 is ejected but here in the backward direction, with a larger translational energy corresponding to a larger collisional exoergicity, here Q_b. Evidently, the product velocity analysis (i.e., the flux-velocity-angle contour map) would be very different for these two situations!

Energy conservation can be expressed

$$E_{total} \equiv E_{tr} + E_{int} - \Delta E_0 = E_{tr}' + E_{int}' = E_{avl} \tag{6.10}$$

where E_{int} and E_{int}' are the internal energies of the reagents and products, respectively, and E_{total} is the maximum 'available energy,' sometimes denoted E_{avl} or simply E. Thus

$$Q \equiv E_{tr}' - E_{tr} = -(\Delta E_0 + \Delta E_{int}) \tag{6.11}$$

where $\Delta E_{int} = E_{int}' - E_{int}$. For $E_{tr}' = 0$, $Q \equiv Q_{min} = -E_{tr}$; for $E_{tr}' = E_{avl}$,

$$Q \equiv Q_{max} = E_{avl} - E_{tr} = E_{int} - \Delta E_0$$

Flux-velocity contour maps may be presented as directly observed in the laboratory frame or as transformed to the c.m. system (see Chapter 4 and MRD), following from the work of K. T. Gillen, A. M. Rulis, and R. B. Bernstein and of D. R. Herschbach and their co-workers. We shall see a number of examples

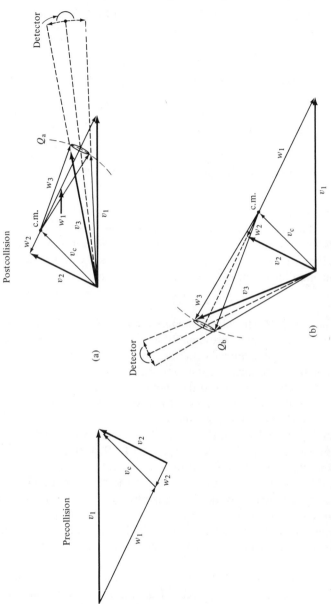

FIG. 6.8. Velocity vector diagrams for a crossed beam reactive scattering experiment. (a) Precollision velocities of Reagents 1 and 2, postcollision velocity of product 3, if scattered predominantly in the forward c.m. direction with a collisional exoergicity Q_a as shown. (b) Postcollision velocity of Product 3 if scattered in the backward c.m. direction with $Q = Q_b \, (> Q_a)$ as shown. The implications with regard to the zone of detection of Product 3 in the laboratory system are obvious from the figure. For details, see reference cited in Fig. 5.7; also MRD, Section 6.2.

of each in the figures which follow. It is important to be aware of the distortion that results from the transformation from the c.m. to the laboratory system. The flux in the laboratory system, say I_{lab}, is related to that in the c.m. system $I_{c.m.}$ by the Jacobian J:

$$I_{lab}(\Theta, v_3) = I_{c.m.}(\theta, w_3) \times J = \frac{v_3^2}{w_3^2} I_{c.m.} \qquad (6.12)$$

(The Jacobian here is the ratio of volume elements in the velocity space; see MRD for further details.)

6.6. Typical experimental results: product angular and velocity distributions

Let us now examine a sampling of experimental results on key reactions studied by the crossed molecular beam scattering technique. First we shall briefly note the results for a prototype of the complex-mode mechanism, namely, the alkali–alkali halide exchange reaction. Figure 6.9 shows a c.m. product flux-velocity contour map for the nearly thermoneutral reaction

$$RbCl + Cs \rightarrow Rb + CsCl \qquad (6.13)$$

which was studied by D. R. Herschbach and co-workers. The reaction proceeds by way of a long-lived complex, presumably RbClCs, evidenced by the near-perfect forward-backward symmetry in the product flux distribution. Not directly apparent from the figure is the fact that the most probable relative velocity of the products is low, and so the energy released into relative translation is not much different from the initial relative kinetic energy, i.e., $Q \approx 0$. A full analysis of the complex-mode reaction mechanism involves consideration of the unimolecular decay of the energized complex; an extension of the RRKM method was developed for this purpose by D. R. Herschbach and co-workers. The theoretical results turn out to be consistent with all the observations on many MX + M' reactions and also for the case for the O + Br$_2$ reaction previously mentioned. With the availability of these and many other experimental studies as a check upon theory, a workable statistical-theoretical model for complex-mode reactions has evolved which can account for the main (observable) features of such flux contour maps. (For further details, see MRD.)

In contrast, the situation for direct-mode reactions is much less subject to generalization. Nevertheless, a great deal of qualitative information can be elicited from the molecular beam data. A good example is the set of reactions

$$H + X_2 \rightarrow HX + X \qquad (6.14)$$

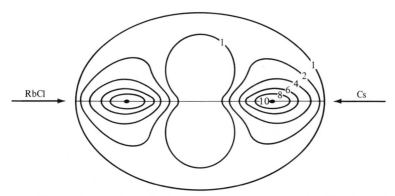

Fig. 6.9. Centre-of-mass flux-velocity contour map of the product(s) of the Cs + RbCl reaction, which proceeds via a long-lived complex. Adapted from unpublished results of W. B. Miller, S. A. Safron, G. A. Fisk, J. D. McDonald, and D. R. Herschbach (1972); see also W. B. Miller, S. A. Safron, and D. R. Herschbach, MIL67, and *J. Chem. Phys.* **56**, 3581 (1972).

where $X \equiv$ Cl, Br, and I. Figure 6.10 shows c.m. product (HX) contour maps for these reactions as derived from experiment. For the reaction of H atoms with Cl_2, the HCl product 'rebounds' directly, with a relatively large recoil velocity. The strongly backward peak in the angular distribution has been interpreted as evidence for a direct encounter with a preferred collinear configuration for the H $\cdots$ Cl $\cdots$ Cl transition state. For the reactions with Br_2 and Cl_2, however, although the angular distributions are strongly anisotropic (implying a direct-mode micromechanism), they show sideward peaking, which suggest bent transition states. Analogous studies of halogen atom–halogen molecule exchange reactions have shown similar behaviour, with the results interpretable in terms of preferred geometry of the trihalogen transition state.

Figure 6.11 is a comparison of flux maps for reactions of K, Rb, and Cs with CCl_4, i.e.,

$$CCl_4 + M \rightarrow MCl + CCl_3 \qquad (6.15)$$

where M is the alkali metal atom. The 'sideward peaking' of the MCl product is noteworthy; it has been interpreted once again in terms of the preferred geometry of the transition state.

Figure 6.12a is a c.m. flux contour map for the SO product of the reaction

$$CS_2 + O \rightarrow SO + CS \qquad (6.16)$$

from the crossed beam measurements of R. Grice et al., carried out at a relatively high collision energy. Here the SO product is strongly 'forward peaked.'

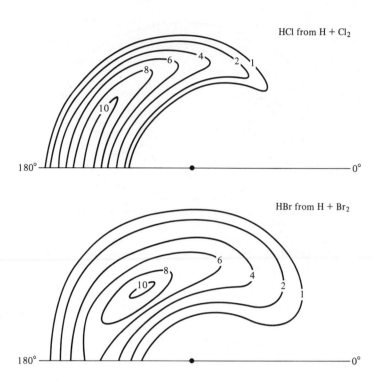

HCl from H + Cl$_2$

HBr from H + Br$_2$

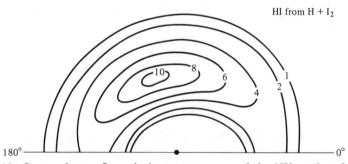

HI from H + I$_2$

FIG. 6.10. Centre-of-mass flux-velocity contour maps of the HX product from the direct-mode H + X$_2$ reactions (X ≡ halogen). The direction of the incident H atom defines 0°, that of the X$_2$ 180°. Adapted from D. R. Herschbach, FAR73, p. 233, based on J. D. McDonald, P. R. LeBreton, Y. T. Lee, and D. R. Herschbach, *J. Chem. Phys.* **56**, 769 (1972); details therein.

Similar behaviour was observed for the analogous reaction,

$$O + Cl_2 \rightarrow ClO + Cl \tag{6.17}$$

as shown in Fig. 6.12b.

Next we shall discuss a few crossed beam reactions in a bit more detail. The alkali–methyl iodide reaction has received considerable attention, since it was the first for which a definitive statement could be made regarding the qualitative shape of the c.m. angular distribution of product. It is, in fact, the classic rebound reaction.

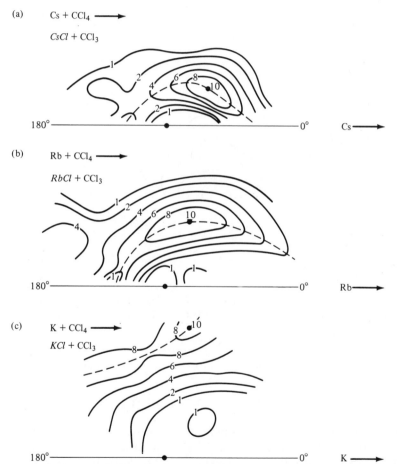

FIG. 6.11. Centre-of-mass flux-velocity contour maps of the MCl product from the crossed beam reactions of $CCl_4 + M$, where M = (a) Cs, (b) Rb, (c) K. Conventions same as Fig. 6.10. Adapted from S. J. Riley, P. E. Siska, and D. R. Herschbach, FAR79, p. 27; details therein.

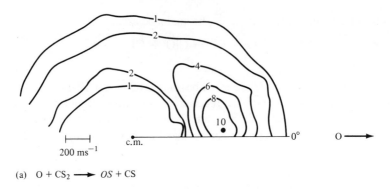

(a) $O + CS_2 \longrightarrow OS + CS$

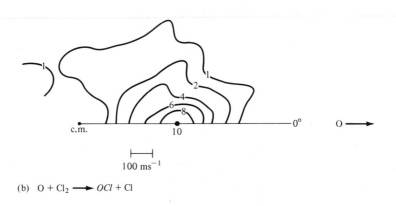

(b) $O + Cl_2 \longrightarrow OCl + Cl$

FIG. 6.12. Centre-of-mass flux-velocity contour maps of the designated products of the O atom reactions with (a) CS_2 and (b) Cl_2. Adapted from R. Grice, GRI79 and references therein.

Figure 6.13 shows the *laboratory* flux contour map for the KI product of the reaction

$$CH_3I + K \rightarrow KI + CH_3 \qquad (6.18)$$

(Note that the locus of the most probable flux is near 85° with respect to the K direction, in the laboratory frame.) Upon conversion to the c.m. system, the KI flux map becomes that displayed in Fig. 6.14, clearly backscattered (peaking at 180° in the c.m. system). The outer dashed circle corresponds to the maximum energetically allowed recoil velocity of the KI with respect to the centre of mass. It is noted that the most probable value of w'_{KI} is approximately

⅘ of its maximum possible value, corresponding to a considerable repulsive energy release in the reaction (see Fig. 4.14). When one integrates over all angles and converts from recoil velocity to final translational energy distribution (using the equations of Chapter 4), it is found that $P(E'_{tr})$ is a sharply peaked function (see Fig. 4.15) with an average value of E'_{tr} some two thirds the value of E_{avl}, i.e., $\langle f'_{tr} \rangle \approx ⅔$. When one integrates over all w'_{KI} the overall angular distribution of KI in the c.m. system, $I(\theta)$ or $d\sigma_R(\theta)/d\omega$, is a smooth function peaked at $\theta = 180°$ which drops to half its maximum at about $90°$.

For comparison, we recall the results already shown in Figs. 4.12 and 4.13 for the reaction of $K + I_2$. Here, the KI product is strongly forward peaked (characteristic of a so-called spectator stripping behaviour, first observed for the $K + Br_2$ reaction by S. Datz and co-workers) and of very low recoil velocity.

6.7. Products' translational energy

Figure 6.15 is a graph of the products' translational energy distribution function $P(E'_{tr})$. The peak in the distribution corresponds to a collision exoergicity $Q \approx 0$. Since such a very small fraction of E_{avl} appears in product recoil, most of it must go into product internal excitation (mainly into the KI, which is found to be highly vibrationally excited). The 'harpoon mechanism' of

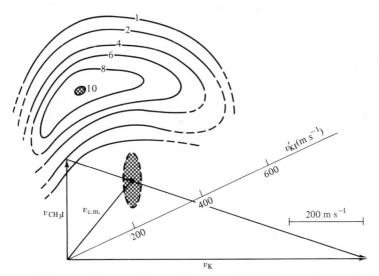

FIG. 6.13. Laboratory flux-velocity contour maps of the KI product from the K + ICH₃ reaction. Adapted from A. M. Rulis and R. B. Bernstein, *J. Chem. Phys.* **57**, 5497 (1972); details therein.

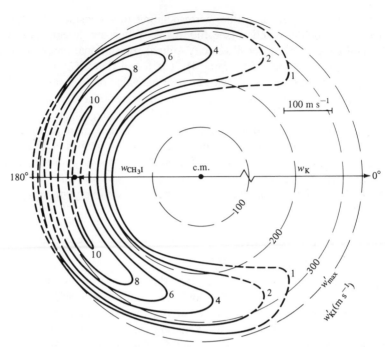

F$_{IG}$. 6.14. Centre-of-mass flux-velocity contour maps of KI product from the K + ICH$_3$ reaction derived from the laboratory map of Fig. 6.13. Adapted from R. B. Bernstein and A. M. Rulis, BER73, and reference of Fig. 6.13.

M. Polanyi for alkali–halogen reactions readily accounts for the main features observed (see MRD).

Returning now to the family of RX + M rebound reactions, studies have been carried out over a wide range of collision energies, by R. B. Bernstein and co-workers R. J. Beuhler, M. E. Gersh, A. M. Rulis, B. E. Wilcomb, H. E. Litvak, S. A. Pace, and H. F. Pang, A. G. Ureña, F. J. Aoiz, and others. The average translational energy of the products $\langle E'_{tr} \rangle$ is found to increase with the reagents' relative translational energy E_{tr}. Figure 6.16 shows results for the CH$_3$I + Rb reaction.

Figure 6.17 shows the dependence of $\langle E'_{tr} \rangle$ upon E_{tr} for four reactions of the type

$$CH_3X + M \rightarrow MX + CH_3 \qquad (6.19)$$

for M $\equiv$ K,Rb and X = Br,I. A theoretical 'impulsive model,' fit to the data, is also shown for each case. For the two CH$_3$Br reactions, experiments have shown clear evidence of a translational energy threshold for reaction, so indicated in Fig. 6.17. (Thresholds for the CH$_3$I reactions, though less directly

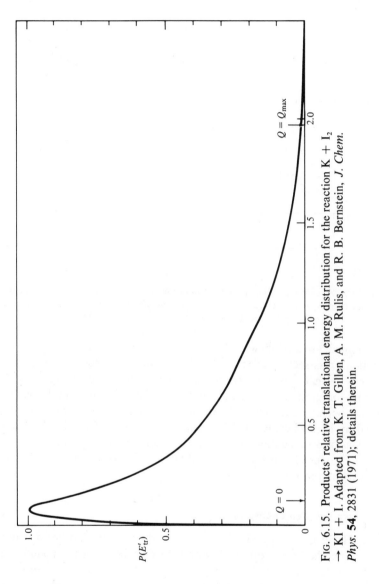

FIG. 6.15. Products' relative translational energy distribution for the reaction $K + I_2$ $\rightarrow KI + I$. Adapted from K. T. Gillen, A. M. Rulis, and R. B. Bernstein, *J. Chem. Phys.* **54**, 2831 (1971); details therein.

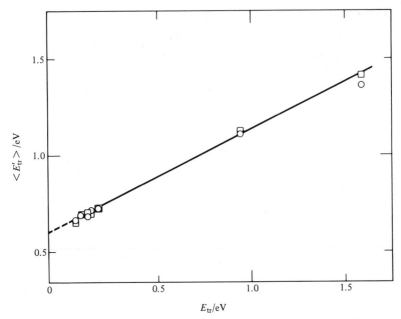

FıG. 6.16. Average products' translational energy as a function of the relative translational energy of the reagents, for the reaction $Rb + ICH_3 \rightarrow RbI + CH_3$. Adapted from A. G. Ureña and R. B. Bernstein, *J. Chem. Phys.* **61**, 4101 (1974).

determined, are also shown.) The translational energy dependence of the total reaction cross section for the K and Rb reactions, shown in Fig. 6.18, is of interest in that no single theoretical model has quite enough flexibility to account for all the experimentally observed features, although the one by B. C. Eu has been frequently used. An information-theoretic approach was introduced by R. D. Levine, H. Kaplan, and co-workers. Classical trajectory calculations on a semiempirical three-body-like potential hypersurface have, however, reproduced most of the qualitative aspects of the overall body of experimental results.

6.8. Two detailed examples

R. R. Herm has reviewed a large body of data on crossed beam reactions of alkali (and alkaline earth) atoms with halogen-containing compounds, pointing out the systematics and discussing the proposed dynamical mechanisms. Many of these reactions are direct mode, but some (e.g., MX + M' exchanges, as mentioned earlier) proceed via the complex mode. One related reaction, studied in considerable detail, which deserves comment here is that involving Hg atoms:

$$Hg + XY \rightarrow HgX + Y \qquad (6.20)$$

where X,Y are halogen atoms. For the endoergic reaction of Hg with I_2, antic-ipating the results of the beam studies of T. M. Mayer, B. E. Wilcomb, and R. B. Bernstein, the minimum energy path can be represented by the potential energy versus reaction coordinate diagram of Fig. 6.19. The reaction goes by

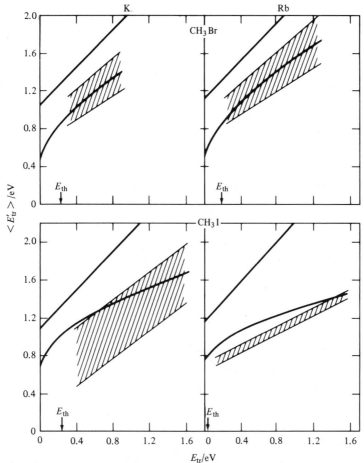

FIG. 6.17. Plots similar to that of Fig. 6.16, for reactions of CH_3Br with K (upper left) and Rb (upper right), and CH_3I with K (lower left) and Rb (lower right). Experimental data indicated by hatched regions. Solid curves theoretical. Diagonal straight lines denote energy conservation limits. Adapted from E. Pollak and R. B. Bernstein, *J. Chem. Phys.* **70**, 3995 (1979) and references therein.

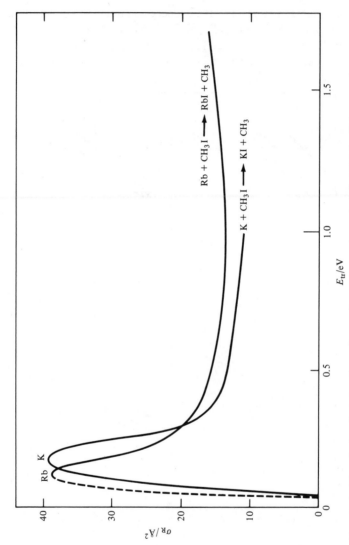

FIG. 6.18. Translational energy dependence of reaction cross sections for $CH_3I + Rb$, $K \rightarrow CH_3 + RbI$, KI. Adapted from K. T. Wu, H. F. Pang, and R. B. Bernstein, *J. Chem. Phys.* **68**, 1064 (1978); details therein.

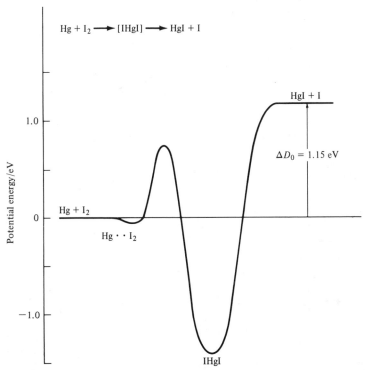

$$\text{Hg} + \text{I}_2 \longrightarrow [\text{IHgI}] \longrightarrow \text{HgI} + \text{I}$$

FIG. 6.19. Potential energy along the reaction coordinate for the $\text{Hg} + \text{I}_2$ reaction, as determined from crossed beam scattering experiments. Note van der Waals minimum, from B. E. Wilcomb, J. A. Haberman, R. W. Bickes, T. M. Mayer, and R. B. Bernstein, *J. Chem. Phys.* **64**, 3501 (1976), followed by a barrier, then a deep 'chemical well,' and the endoergic rise to $\text{HgI} + \text{I}$ product level. Adapted from T. M. Mayer, B. E. Wilcomb, and R. B. Bernstein, *J. Chem. Phys.* **67**, 3507 (1977); details therein.

way of a long-lived complex (presumed to be IHgI); however, unless the total energy exceeds the threshold of 1.15 eV, the complex can only decay back to reagents. If the available energy exceeds 1.54 eV, the process of collision-induced dissociation (CID) can occur, i.e.,

$$\text{Hg} + \text{I}_2 \rightarrow \text{Hg} + \text{I} + \text{I} \qquad (6.21)$$

This process may be direct (a ballistic mechanism, as would result if Hg were replaced by Xe, for example), or indirect, by the formation of a temporarily super-excited HgI, say $\text{HgI}^{\dagger\dagger}$ with an internal energy in excess of the 0.39-eV bond dissociation energy of HgI, i.e.,

$$\text{Hg} + \text{I}_2 \rightarrow \text{HgI}^{\dagger\dagger} + \text{I} + \text{I} \qquad (6.22)$$

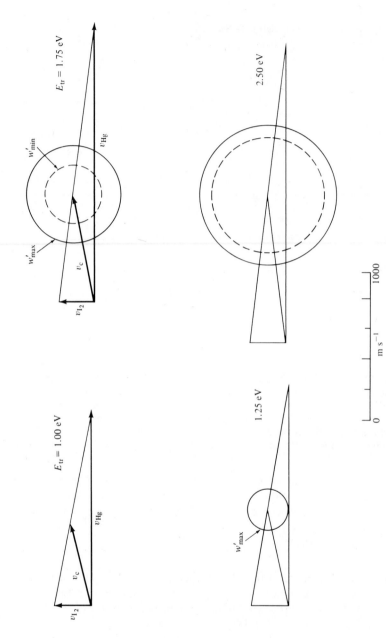

FIG. 6.20. Newton diagrams defining the conditions of the Hg + I$_2$ crossed beam reaction experiments of Mayer et al., reference of Fig 6.19; details therein. Detected product: HgI.

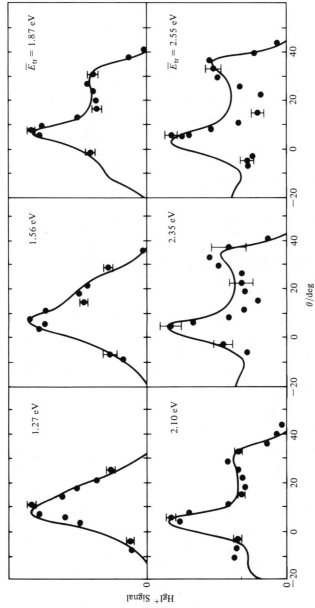

FIG. 6.21. Laboratory angular distributions of HgI product from the Hg + I₂ reaction at the indicated collision energies. Points, experimental; curves, theoretical fits assuming long-lived complex. Adapted from Mayer et al., reference of Fig. 6.19; details therein.

Figure 6.20 shows representative Newton diagrams characterising the experiments. The maximum allowed recoil velocities of HgI (relative to the c.m.) are denoted w'_{max}. For $E_{tr} < E_{th}(\text{HgI}) = 1.15$ eV, no reaction can occur; for $E_{tr} > E_{th}(\text{CID}) = 1.54$ eV, there is a minimum value of w' below which E'_{tr} is too small, i.e., E'_{int} too large ($> D_0(\text{HgI}) = 0.39$ eV) for HgI to survive as a stable, detectable product. Thus there is a sphere in product velocity space within which HgI is 'forbidden,' shown in projection on the beam plane as the dashed circle. A velocity analysis of the HgI product at given laboratory angle would be expected to show a zero signal level over an appreciable range of speeds (the chord through the inner sphere). An angular distribution of HgI could therefore show a bimodal distribution with a minimum near the centroid angle.

Figure 6.21 displays observed laboratory angular distributions of HgI product at several E_{tr}, together with computer simulations based upon a statistical-dynamic model assuming a long-lived (IHgI) complex, for which the c.m. flux contour map at $E_{tr} = 1.9$ eV is that shown in Fig. 6.22. The model predicts energy-dependent total cross sections for IHgI formation, HgI formation, and CID, as displayed in Fig. 6.23. In the region of the threshold for HgI forma-

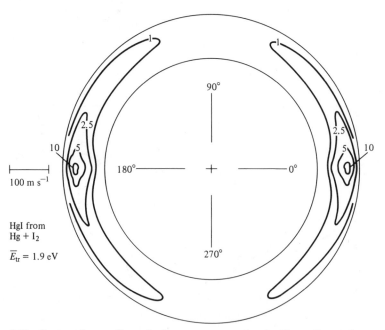

FIG. 6.22. Centre-of-mass flux-velocity contour map for the Hg + I_2 reaction at the indicated energy. Adapted from BER78 based on the results of Mayer et al., reference of Fig. 6.19.

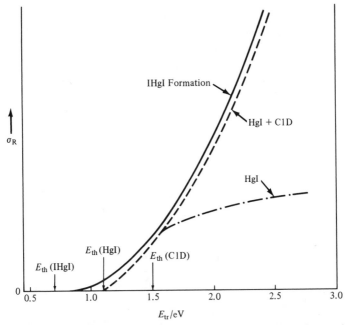

FIG. 6.23. Theoretical estimation of the translational energy dependence of various cross sections for the Hg + I$_2$ reaction. The solid curve denotes the cross section for IHgI (complex) formation, the dashed curve is for HgI formation; above 1.5 eV, collision-induced dissociation (CID) sets in. Adapted from Mayer et al., reference of Fig. 6.19; details therein.

tion, convoluting a line-of-centres cross section function (Eqn. 6.8) together with the known spread in E_{tr} allows one to predict the in-plane yield of HgI as a function of E_{tr}, as shown in Fig. 6.24. The calculation compares well with the measured points.

On the basis of all the experimental crossed beam results, and taking cognizance of the known potential functions for the diatomics (I$_2$ and HgI) and the triatomic (IHgI) complex, it has been possible to estimate the adiabatic potential hypersurface for the Hg + I$_2$ reaction system (also consistent, of course, with the minimum energy path of Fig. 6.19), shown in Fig. 6.25. The lower right side of the diagram describes the formation of the IHgI complex via insertion of an Hg atom into an I$_2$ molecule; the upper left describes the decay of the complex to HgI product.

Next we shall briefly examine the results of an elegant state-of-the-art crossed beam study of the 'bellwether reaction,'

$$ F + H_2 \rightarrow HF + H \tag{6.23} $$

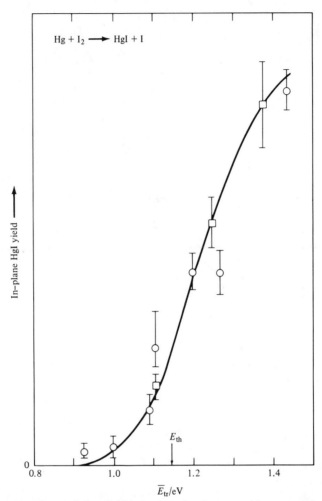

Fig. 6.24. Experimental threshold behaviour for the crossed beam reaction of Hg + I_2. Solid curve is best-fit (convoluted) product yield function, corresponding to the indicated threshold. Adapted from B. E. Wilcomb, T. M. Mayer, R. B. Bernstein, and R. W. Bickes, *J. Amer. Chem. Soc.* **98**, 4676 (1976) and reference of Fig. 6.19; details therein.

(and its isotopic analogue with D_2) by Y. T. Lee et al. Figure 6.26a shows a c.m. map of the HF product distribution at $\overline{E}_{tr} = 13.3$ kJ mol^{-1}, superimposed on the Newton triangle for the experiment. Spheres in recoil velocity space correspond to maximum velocities (relative to the c.m.) for HF in rotationless states $v' = 1, 2,$ and 3 ($v' = 0$ is omitted since virtually no HF is formed in the ground vibrational state). The predominance of backscattered product in $v' = 3$ is to be noted. Figure 6.26b is a similar map for DF from the D_2 reaction at the indicated collision energy. Since DF has smaller vibrational spacings, there are more vibrational states available (for the same E_{avl}). Clear evidence for the population by reaction of resolved states $v' = 2, 3$, and 4 is noted. These data are complementary to results on population inversion in these reactions as measured by IR chemiluminescence or chemical laser experiments. The significant contribution of the crossed beam results is the quantitative information on the *angular* as well as state distribution of the reaction product.

One especially interesting feature of the data of Fig. 6.26 is the 'lobe structure' in the $v' = 2$ angular distribution. This was found only over a small energy range near the indicated value of E_{tr}. *Ab initio* theoretical computations by R. E. Wyatt and co-workers have predicted certain resonance behaviour which can be closely related to this observation (i.e., the maximum in the c.m.

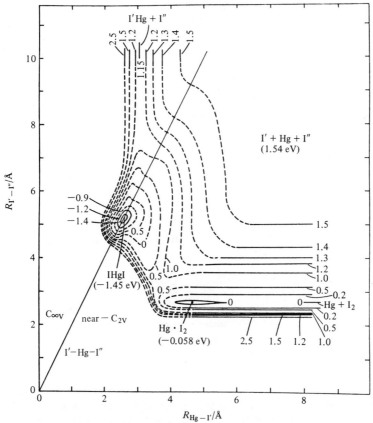

FIG. 6.25. Empirically derived potential energy surface appropriate to the Hg + I$_2$ reaction. Adapted from T. M. Mayer, J. T. Muckerman, B. E. Wilcomb, and R. B. Bernstein, *J. Chem. Phys.* **67**, 3522 (1977); details therein.

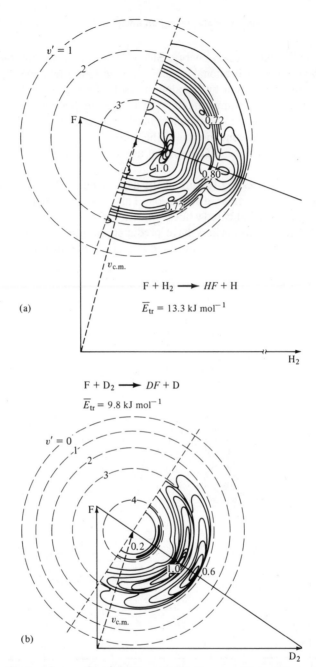

FIG. 6.26. Centre-of-mass flux-velocity contour map, superimposed on the nominal Newton diagram, for (a) HF from the $F + H_2$ reaction at a collision energy of 0.14 eV and (b) DF from the $F + D_2$ reaction at $\overline{E}_{tr} = 0.10$ eV. Note the 'bimodal' angular distribution for $v'(HF) = 2$ in (a) compared with the simple maximum at the back-

differential reaction cross section for the $v' = 2$ state shifting from $180°$ to smaller angles at certain critical energies). Because of the intense theoretical and experimental work on Reaction 6.23, it promises to be something of a landmark in the history of molecular beam kinetics.

6.9. Orientation effect on reaction probability

One final topic deserves mention in this chapter: the influence of relative orientation of reagents upon their reactive cross sections, i.e., the steric effect in elementary, direct-mode chemical reactions. As discussed in Chapter 3, it is possible to state-select and then orient polar molecules (e.g., CH_3I) in a molecular beam using suitable electric fields and then allow them to collide with a crossed beam of a co-reagent (e.g., K or Rb), with the dipole moment of the molecule either parallel or antiparallel to the incident relative velocity vector. In one orientation (the favourable one, say f) the reaction cross section is large, whereas in the other (the unfavourable, u) configuration, steric hinderance can serve to reduce the reaction probability. Such experiments were first reported by P. R. Brooks and E. M. Jones and by R. J. Beuhler, R. B. Bernstein, and K. H. Kramer. Figure 6.27 shows a set of experimental results for two crossed beam, oriented molecule reactions. The top portion of Fig. 6.27 refers to the reaction

$$K + Cl_3CH \rightarrow KCl + CCl_2H \qquad (6.24)$$

involving the nearly spherical chloroform molecule. No asymmetry, i.e., no orientation dependence, of the laboratory angular distribution is observed. However, for the analogous reaction of methyl iodide (Reaction 6.18), a gross orientation effect is noted, namely, the reactivity for the unfavourable orientation (u) is less than half that for the favourable one (f). More accurate experiments for the closely related reaction

$$Rb + ICH_3 \rightarrow RbI + CH_3 \qquad (6.25)$$

have shown that for the u orientation there is essentially *no* detectable reaction, provided the angle between the dipole moment and the relative velocity is less that about $45°$. For the u arrangement, the reactive cross section for product backscattering in the c.m. system is <10 percent of that for the f orientation.

Clearly, the crossed molecular beam technique has contributed a great deal of new, detailed information on the microscopic mechanism and molecular dynamics of elementary chemical reactions. All of the material discussed in

scattering angle for all other cases. Adapted from R. K. Sparks, C. C. Hayden, K. Shobatake, D. M. Neumark, and Y. T. Lee, *Proc. 3rd Intl. Quantum Chem. Conf.,* Kyoto, Japan (1979); details therein.

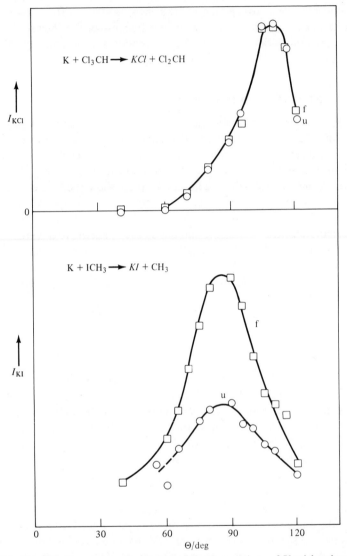

FIG. 6.27. Results of experiments on crossed beam reactions of K with oriented molecules of CCl_3H (top) and CH_3I (bottom). Laboratory angular distributions of the KI product from the $K + ICH_3$ reaction show directly a strong orientation effect: f denotes favourable orientation (K–ICH$_3$), u unfavourable (K–H$_3$CI). No such asymmetry is observed for the CCl_3H reaction. Adapted from G. Marcelin and P. R. Brooks, FAR73, p. 318.

this chapter refers to reactions of ground-state reagents on adiabatic potential surfaces. The role of excited states of reactants upon the reaction cross sections and dynamics is the subject of the next chapter.

Review literature

Reactive scattering and mainly experimental
HER62, ROS62, FIT63, HAS64, MCD64, MCD64x, HER65, PAU65, ZOR65, GRE66, HER66, MUS66, ROS66, PAU68, ROS69, SCH69, ROS70, SCH70, WOL70, BER71, MAS71, DUB72, KIN72, LEE72, MCC72, BER73, BER73a, DIN73, DUB73, FAR73, FLU73, HER73, HER73a, PAR73, WON73, FAR74, TOE74, BER75a, GRI75, LAW75, PAU75, REU75, BRO76, HER76, BER77, BRO77, FAR77, GEI77, HER77, BER78, BER78a, ZEW78, BER79, BER79a, BER79b, DAV79, FAR79, GRI79, HER79, KWE79, LEV79x, SCH79, SMI79, STO79, ZAR79, LEE80, SMI80, ZAR80

Potential surfaces and mainly theoretical
ROS65, MIL67, LEV69, LES70, LIG71, LEV72, SCH72, SMI72, GEO73, LEV73, CHI74, EYR74, LEV74, NIK74, POR74, BAL75, BAL75a, BER75, EYR75, MIC75, KUN76, LES76, MIL76, POR76, TRU76, TUL76, BAL77, TRU77, WYA77, GLO79, KUN79, LIG79, TRU79, TRU79a, WYA79, EYR80, HAS81, HEN81, KUP81, SCH81, TRU81

7 Relative efficacy of vibrational, rotational, and translational energy in promoting elementary reactions

7.1. Measurement of state-to-state rates

In Chapter 2, concerned with state-to-state reaction cross sections, there was presented a theoretical framework for dealing with the role of the reagents' internal excitation upon reactivity, and the relative importance of internal energy versus translational energy. In this chapter, we shall have a look at a number of experimental results bearing strongly on this subject. Considerable information on the influence of vibrational excitation on reactivity has come from bulk gas-phase experiments, e.g., those of J. Wolfrum et al., J. H. Birely and co-workers, S. H. Bauer and co-workers, and others, involving selective laser excitation of reagents. Additional results of great importance have been obtained by a flow system technique devised by J. C. Polanyi and co-workers. A certain well-chosen reaction occurs in a so-called pre-reactor to form molecules in an ensemble of highly excited vibrational states which themselves assume the role of reagents in a bimolecular reaction downstream. The overall vibrational enhancement in rate constant (as measured by any of several techniques) can be deconvoluted to yield a set of state-dependent rate coefficients.

Schematically,

$$B + CD \overset{a}{\to} BC^\dagger + D \tag{7.1a}$$

$$A + BC^\dagger \overset{b}{\to} AB + C \tag{7.1b}$$

Given a knowledge of the vibrational state population in BC from the pre-reaction shown in Eqn. 7.1a and a measurement of the rate of formation of AB for several different ensembles of reagents BC from different pre-reactions, one can deduce the state-dependence of the rate of Reaction 7.1b.

Obviously, there must be at least as many different ensembles (sets of reagent states) as rate coefficients k_v to be evaluated (for Reaction 7.1b). We can spell it out a little more explicitly, as follows:

$$A + BC(v) \overset{k_{v'v}}{\to} AB(v') + C \tag{7.2a}$$

$$A + BC(v) \overset{k_v}{\to} AB(\text{all } v') + C \tag{7.2b}$$

where $k_v = \sum_v k_{v'v}$ is one of the desired rate coefficients for Reaction 7.1b.

Then for the ith ensemble,

$$n_{AB}^{(i)} = n_A^{(i)} \sum_v k_v n_{BC(v)}^{(i)} = n_A^{(i)} n_{BC}^{(i)} \sum_v k_v f_v^{(i)} \equiv k^{(i)} n_A^{(i)} n_{BC}^{(i)} \qquad (7.3)$$

Here

$$k^{(i)} \equiv \sum_v k_v f_v^{(i)} \qquad (7.4)$$

is the measured overall bimolecular rate constant for the ith ensemble. Knowing the relative vibrational populations of AB from Reaction 7.2a for each of the i pre-reactions, i.e., the set of $f_v^{(i)}$, and measured values of $k^{(i)}$, one can deduce the required set of k_v's for Reaction 7.2b. This procedure was first carried out by B. A. Blackwell, J. C. Polanyi, and J. J. Sloan.

Obviously, it would be even more desirable to measure the state-to-state rate constants of Eqn. 7.2a than simply k_v, the rate constant *out* of state v, that of Eqn. 7.2b. This has also been done, using IR chemiluminescence from AB(v'); results will be shown later.

7.2. Results showing vibrational enhancement of reactivity: bulk gas-phase experiments

Illustrating the early results obtained with the pre-reactor method is Fig. 7.1, in which k_v is plotted versus v (from 0 to 4) for the reaction

$$Br + HCl(v) \rightarrow HBr + Cl \qquad (7.5)$$

for which $\Delta E_0 = 69$ kJ mol^{-1}. The sharp increase in k_v for $v \geq 2$ (E_{int} just in excess of the endoergicity) is to be noted. Figure 7.2 is a schematic drawing of the minimum energy path for this reaction, showing the zero-point level of HCl and the first two excited vibrational states, as well as the zero-point level of the HBr product.

Figure 7.3 shows results for three related endoergic reactions, plotted as relative rate constants k_v versus the reagent's vibrational energy, expressed as a fraction of the energy barrier E_b for the reaction (values listed). Once again, a 'vibrational threshold' for reaction is observed, in good accord with expectation.

The reaction shown in Eqn. 7.5 has also been investigated by D. Arnoldi and J. Wolfrum, who used a somewhat more direct laser excitation technique that yielded values for k_2 and k_1 relative to k_0. An enhancement of 11 orders of magnitude (k_2/k_0) was deduced. Table 7.1 lists approximate values of k_v (near 298 K) for $v = 0$ to 4 for the Br + HCl reaction.

Referring back to Chapter 2, the empirical equation 2.32 implies that

$$\ln(k_v/k_0) = -(E_a - \alpha E_v)/kT \qquad (7.6)$$

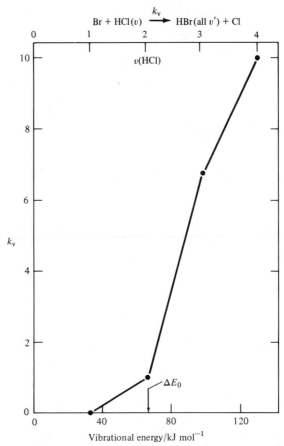

FIG. 7.1. Dependence of the rate of the Br + HCl(v) reaction upon vibrational state v (and thus vibrational energy) of the HCl reagent. The arrow designates the endoergicity, ΔE_0. Adapted from D. J. Douglas, J. C. Polanyi, and J. J. Sloan, *J. Chem. Phys.* **59**, 6679 (1973); details therein.

where α is the fraction of the vibrational energy of the reagent 'utilized' to reduce the activation energy E_a of the reaction, so that the vibrational enhancement at a given temperature could be plotted as ln k_v versus E_v with a slope

$$\left(\frac{\partial \ln k_v}{\partial E_v} \right)_T = \alpha / kT \qquad (7.7)$$

(i.e., Eqn. 2.33). J. H. Birely and J. L. Lyman (BIR75) reviewed the early laser-excitation results and found that for reactions of vibrationally excited H_2, HCl, OH, and O_3, the empirical values of α were usually less than unity, but not simply correlated with E_a or ΔE_0.

TABLE 7.1

Bimolecular rate coefficients k_v for Reaction 7.5 (from BER77)

v	0	1	2	3	4
k_v (M^{-1}s^{-1})	1×10^{-2}	2×10^4	1×10^9	7×10^9	1×10^{10}

Another elementary reaction investigated by the J. C. Polanyi group using the pre-reactor technique is the reaction of hydroxyl with chlorine atoms:

$$OH(v) + Cl \rightarrow HCl(v') + O \qquad (7.8)$$

which is nearly thermoneutral for $v = v' = 0$. Two different pre-reactions were used:

$$H + O_3 \rightarrow OH(v = 6 \text{ to } 9) + O_2 \qquad (7.9a)$$

$$H + NO_2 \rightarrow OH(v = 1 \text{ to } 3) + NO \qquad (7.9b)$$

Results for Reaction 7.8 are plotted in Figure 7.4 as k_v versus v (for $v = 1$ to 3 and 7 to 9).

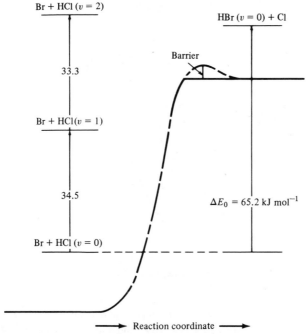

FIG. 7.2. Energy level diagram superimposed on schematic reaction coordinate for the Br + HCl(v) reaction of Fig. 7.1. Adapted from D. Arnoldi and J. Wolfrum, *Ber. Bunsenges. Phys. Chem.* **80**, 892 (1976); see also BER77.

Additional information was obtained regarding the vibrational state distribution of the HCl product of Reaction 7.8. As shown in Fig. 7.5, the vibrational states v' of the HCl product formed from different v states of the OH reagent were very well correlated with the vibrational states of the reagent. Thus the reagent's vibrational energy was, in effect, transformed into product's vibrational energy, i.e., there was found to be a proclivity for the state-to-state reaction to preserve its thermoneutrality: $\Delta E_v \approx \Delta E_{v'}$. One can explain this behaviour if the H atom transfer occurs at large impact parameters and the high-amplitude vibrating OH forms an 'extended' $O \cdots H \cdots Cl$ transition state or activated complex which decays to yield vibrationally excited HCl. It is noteworthy that (irrespective of the explanation) the existence of this pseudoresonant vibrational energy transfer process has made it possible to produce a stream of HCl molecules with vibrational energies of about 3 eV, through the combination of Pre-reaction 7.9a and Reaction 7.8. This approach has

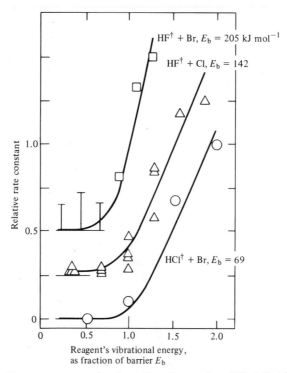

FIG. 7.3. Specific rate constants for the endoergic reactions HF + Br, Cl and HCl + Br as a function of the vibrational excitation of the reagent molecule, expressed as a fraction of the barrier height, E_b. Successive ordinate scales displaced. Adapted from D. J. Douglas, J. C. Polanyi, and J. J. Sloan, *Chem. Phys.* **13**, 15 (1976); details therein.

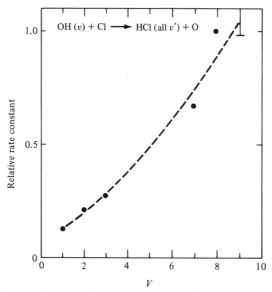

FIG. 7.4. Relative rate constant for the thermoneutral reaction OH + Cl → HCl + O as a function of the vibrational state of the OH reagent. Adapted from B. A. Black-well, J. C. Polanyi, and J. J. Sloan, *Chem. Phys.* **24**, 25 (1977); details therein.

practical implications with regard to the design of hydrogen halide chemical lasers (cf. Chapter 3).

7.3. Theoretical framework for vibrational enhancement

A theoretical approach to this subject is the use of a statistical (i.e., density of states) model to calculate the prior-expectation value of the rate constant, say $k_v^0(T)$, by R. D. Levine and J. Manz. For the simple case of an atom–diatomic molecule exchange reaction, in the limit of the rigid-rotor, harmonic oscillator (RRHO) approximation, as formulated by J. L. Kinsey, it is possible to obtain a closed-form expression for the 'prior' rate coefficient $k_v^0(T)$ as a function of the reduced energy $(E_v - \Delta E_0)/kT$. Here, E_v is the vibrational energy of the reagent diatomic (above its zero-point level). This result is shown in Fig. 7.6. In the endoergic limit, i.e., for $E_v - \Delta E_0 < 0$, it is found that

$$k_v^0(T) \propto \exp[(E_v - \Delta E_0)/kT] \tag{7.10}$$

so that the slope of the plot of $\ln k_v^0(T)$ in Fig. 7.4 approaches unity (in this limit). In the exoergic limit, i.e., $E_v - \Delta E_0 > 0$, $k_v^0(T)$ increases more slowly with E_vib according to a simple power law:

$$k_v^0(T) \propto [(E_v - \Delta E_0)/kT]^{5/2} \tag{7.11}$$

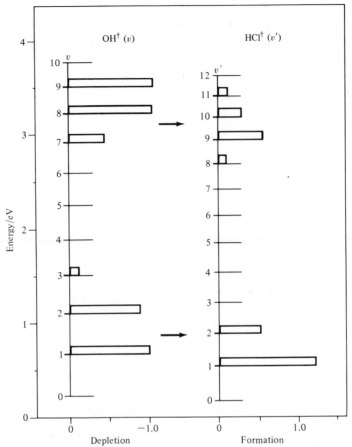

FIG. 7.5. State-to-state rate data on the reaction of Fig. 7.4. Experimental relationship between vibrational energy of reagent OH and that of product HCl. Reagent vibrational energy is effectively 'transposed' into product vibrational energy. Adapted from reference of Fig. 7.4.

The prior-expectation behaviour of Fig. 7.6 is well confirmed by the experimental results for the reaction described in Eqn. 7.5 (listed in Table 7.1), plotted in Fig. 7.7. Here, the ordinate is ln k_v(298 K), the abscissa E_{vib}. In the 'endo' region, the slope is essentially that expected from Eqn. 7.10, i.e., $(kT)^{-1}$, with $T = 298$ K. It should be noted that a semilog plot (such as Figs. 7.6 and 7.7) emphasizes the very strong, 'order of magnitude' vibrational enhancements in the endo region, below the 'vibrational threshold' ΔE_0 (or, better, E_a). However, a simple linear plot (such as Figs. 7.1 and 7.3) is more instructive in discussing the rather small 'post-threshold' vibrational enhancements.

The empirical enhancement equation 7.6 reduces to the prior expectation

(Eqn. 7.10) in the endo limit, provided we take $\alpha = 1$ (i.e., *all* the vibrational excitation energy of the reagent is used to overcome the activation energy of the reaction).

There have been a large number of experimental studies on vibrational enhancement for reactions of atoms and simple molecules, including the following:

$$Na, Ba, Br, Cl, O + HF^\dagger \rightarrow F + HZ \qquad (Z \equiv M, X, O)$$

$$H, Na, K, O + HCl^\dagger \rightarrow Cl + HZ \qquad (Z \equiv H, M, O)$$

$$Cl + OH^\dagger \rightarrow O + HCl \qquad\qquad\qquad (7.12)$$

$$O + CN^\dagger \rightarrow CO + N$$

$$O_3 + NO^\dagger \rightarrow O + NO_2$$

(See the review articles, e.g., BER77, for references and discussion.)

For polyatomic reagents, the situation is, of course, more complicated.

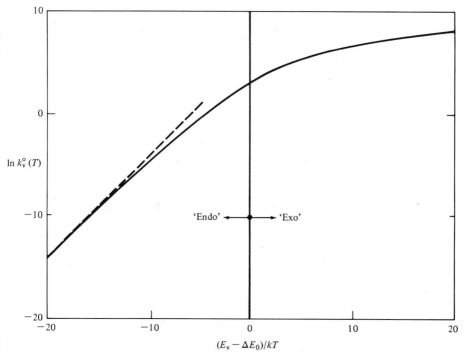

FIG. 7.6. Prior-expectation dependence of the rate constant for an endoergic atom–diatomic molecule reaction upon vibrational energy of the reagent. Semilog plot of $k_v^0(T)$ versus the 'reduced available energy' $(E_v - E_0)/kT$, for the RRHO model. Adapted from R. D. Levine and J. Manz, *J. Chem. Phys.* **63**, 4280 (1975); see also BER77.

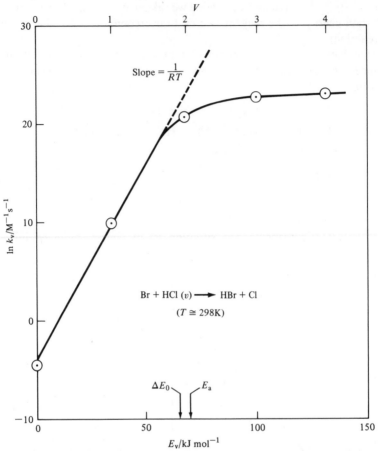

FIG. 7.7. Semilog plot of the rate constant for the Br + HCl reaction as a function of the vibrational energy of the reagent (cf. Fig. 7.6). The dashed reference line with a slope of $1/kT$ corresponds to the prior expectation for small values of the reduced available energy (i.e., the unit slope of Fig. 7.6). Adapted from BER79b based on data of J. C. Polanyi and J. Wolfrum from references of Figs. 7.1, 7.2, and 7.3; see also BER77.

Nevertheless, in a recent study of the bulk gas-phase reaction

$$SF_6^\dagger + Na \rightarrow NaF + SF_5 \qquad E_a \approx 14 \text{ kJ mol}^{-1} \qquad (7.13)$$

by F. R. Grabiner and co-workers M. Eyal and V. Agam using a CO_2 laser for vibrational excitation of the SF_6, a plot of $\ln k_v$ versus E_v at constant temperature was essentially linear and $(\partial \ln k_v / \partial E_v)_T$ yielded $\alpha = 0.38$. From an experimental viewpoint, it is of interest to note that an internal excitation corresponding to the absorption of *more* than one photon (10.9 μm) per SF_6 molecule would be required to overcome the activation energy of Reaction 7.13, whereas only about 0.1 photon per molecule was actually absorbed under the conditions of the above experiment.

Before leaving the subject of bulk gas-phase laser excitation experiments on vibrational enhancement of reactivity, it is worthwhile to mention the recent result of M. Kneba and J. Wolfrum on the simplest of all the elementary reactions, the $H + H_2$ exchange. The actual system studied was $D + H_2$:

$$D + H_2(v = 1) \rightarrow HD + D \tag{7.14}$$

for which a potential energy diagram is depicted in Fig. 7.8. Ground-state H_2 was excited to the state $v = 1$ by collision with $HF(v = 1)$, prepared by irradiation from a resonant HF chemical laser. Deuterium atoms from a microwave discharge of D_2 were allowed to flow into the H_2–HF–He carrier gas stream and react with $H_2(v = 1)$ to yield H atoms, detected by Lyman-α absorption. After careful analysis of the data, the authors deduced a value for the bimolecular rate coefficient for Reaction 7.14 of about 10^{11} cm^3s^{-1} at 298 K. Thus $k_{v=1}(298)$ is some 4×10^4 times larger than the known value of $k_{v=0}(298)$ for the reaction ground-state H_2. This enhancement is much larger than expected on the basis of quasiclassical trajector simulations, and so there is room for more theoretical computational work (even for the H_3 system!). For a discussion of this and other bimolecular, vibrationally excited reactions, see KNE80.

7.4. Experimental results: molecular beam experiments

Now we turn to the molecular beam approach. Clearly, we stand to gain considerable new dynamic information because the experiments can be carried out with separate control of the relative translational energy as well as the reagents' electronic, vibrational, and rotational energy. In addition, one can measure the products' velocity-angle distribution. By means of laser-induced fluorescence, products' internal state populations can be determined. It has also been possible to measure the degree of polarization of the products' rotational angular momenta. The crossed molecular beam approach also permits direct observation of the dependence of the reactivity upon the reagents' mutual orientation.

The first crossed beam experiment on vibrational enhancement was that of T. J. Odiorne, P. R. Brooks, and J. V. Kasper. The reaction studied was

$$HCl(v) + K \rightarrow KCl + H \tag{7.15}$$

For the ground-state reagent, $HCl(v = 0)$, the reaction is slightly endoergic. As shown by P. R. Brooks and co-workers, the total reaction cross section

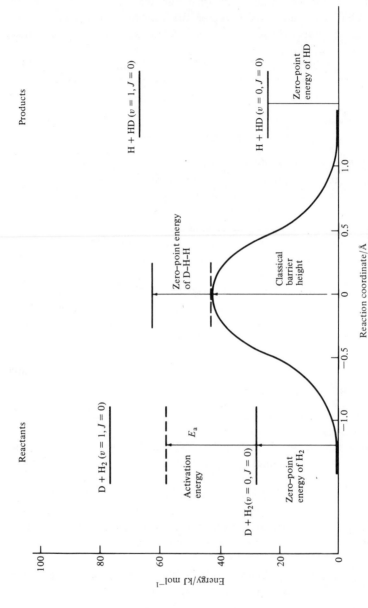

FIG. 7.8. Energy level diagram superimposed on reaction coordinate based on *ab initio* potential surface for H₃, for the collinear configuration, appropriate for the D + H₂(v = 0,1) reaction. Adapted from M. Kneba, U. Wellhausen, and J. Wolfrum, *Ber. Bunsenges. Phys. Chem.* **83**, 940 (1979); details therein.

$\sigma_R(E_{tr})$ shows a strong, positive translational energy dependence immediately post-threshold (about 6 kJ mol^{-1}). However, upon laser excitation of the HCl to the $v = 1$ state (at low E_{tr}), a large increase in KI yield (some two orders of magnitude) was observed (Fig. 7.9). This was the case even when E_{tr} was increased to values well in excess of the 34.5 kJ mol^{-1} corresponding to the 0 $\rightarrow$ 1 vibrational excitation energy of HCl.

The translational energy range was great enough to allow comparison of

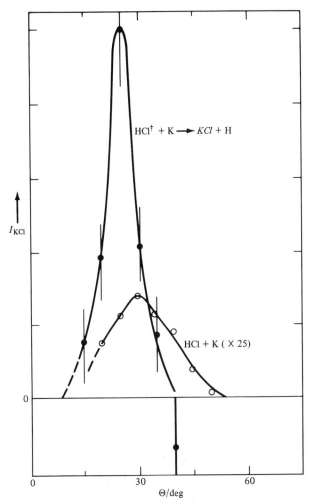

FIG. 7.9. Influence of laser excitation of HCl upon its reactivity with K in a crossed beam scattering experiment. Adapted from T. J. Odiorne, P. R. Brooks, and J. V. Kasper, *J. Chem. Phys.* **55**, 1980 (1971); details therein.

results at constant total energy. It was found that the efficacy of the reagent's vibrational energy was far greater than that of the relative translational energy in promoting reaction. In fact, for $E_{tr} \geq 34$ kJ mol^{-1}, $\sigma_R(E_{tr})$ passes over a maximum and declines with E_{tr}.

In further experiments, P. R. Brooks and co-workers have determined the relative cross sections for different rotational states of HCl($v = 1$) in Reaction 7.15. Using a grating-tuned HCl chemical laser, they were able to carry out selective excitation of specific v,J states of the HCl molecular beam (prior to collision).

Figure 7.10 shows the rotational dependence of the reaction cross section with $v = 1$ HCl reagent. To be noted is the strong negative J dependence (at fixed E_{tr}). Without comparison data for HCl ($v = 0,J$), it is premature to discuss possible dynamic models to explain the qualitative aspects of the J dependence of Fig. 7.10.

7.5. A detailed crossed beam study: the MF + K exchange reaction

Crossed beam experiments on the rotational energy dependence of both differential and integral reaction cross sections have been carried out by R. B. Bernstein and co-workers S. Stolte, A. E. Proctor, W. M. Pope, and L. Zandee for the reactions:

$$CsF(J) + K \rightarrow Cs + KF \quad \text{(endoergic)} \quad (7.16a)$$

and

$$RbF(J) + K \rightarrow Rb + KF \quad \text{(exoergic)} \quad (7.16b)$$

These are members of the so-called complex-mode MX + M' reactions, proceeding by way of a long-lived ($\tau \geq 1$ ps) MXM' intermediate, as discussed in Chapter 6.

The influence of the (average) rotational energy of the alkali fluoride molecule upon its reactivity with K was measured using velocity-selected MF beams. The average relative translational energy E_{tr} ranged from 13 to 26 kJ mol^{-1} for the CsF reaction and from 13 to 20 kJ mol^{-1} for RbF. A quadrupole focusing field rotational state selector was used to provide low-J beams (rotational energies $E_{rot} \leq 1$ kJ mol^{-1}), whose reactivity could be compared with rotationally Boltzmann MF beams, $E_{rot} \approx 10$ kJ mol^{-1}. Thus a difference of some 9 kJ mol^{-1} in average rotational energy of the alkali halide reagent is available. The state-selection technique was discussed in Chapter 3; the apparatus for this study is sketched in Fig. 3.11.

For each of the reactions 7.16, the reactive/nonreactive branching ratio for the decay of the MFM' complex was first measured as a function of $\overline{E}_{tr}$ using velocity-selected, but internally Boltzmann, beams of the alkali fluorides. As will be seen, the reactive branching fractions $F_R(\overline{E}_{tr})$ showed the expected behaviour, i.e., positive E_{tr} dependence of F_R for the endoergic Reaction 7.16a,

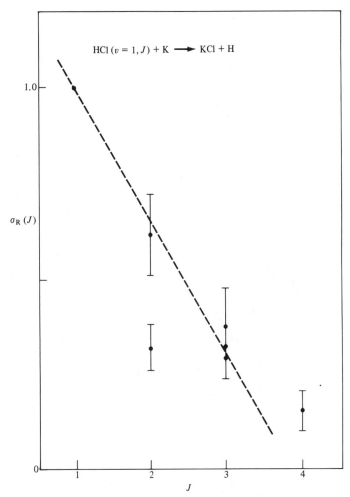

Fig. 7.10. Dependence upon rotational state J of the cross section for the reaction of $HCl(v = 1, J)$ with K, as determined from a crossed beam scattering experiment in which the HCl beam was selectively excited by various lines from a resonant HCl chemical laser. Adapted from H. H. Dispert, M. W. Geis, and P. R. Brooks, *J. Chem. Phys.* **70**, 5317 (1979); details therein.

negative for the exoergic 7.16b. Then the influence upon F_R of the removal of the approximately 9 kJ mol^{-1} of rotational energy from the MF reagent molecules at fixed $\overline{E}_{tr}$ was measured, yielding the desired 'rotational energy effect,' to be discussed later.

Let us now have a detailed look at the body of experimental results on Reactions 7.16 which have made it possible to evaluate the rotational energy effect upon the reactivity of these alkali halides. First we should note the kine-

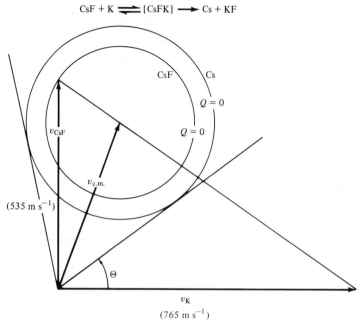

$$CsF + K \rightleftharpoons [CsFK] \longrightarrow Cs + KF$$

FIG. 7.11. Newton triangle appropriate for the crossed beam reaction CsF + K → Cs + KF at a collision energy of 14 kJ mol^{-1}. The nominal angular range of nonreactive scattering of the CsF is governed by the $Q = 0$ circle labelled CsF. For the reactive scattering, since the most probable collisional exoergicity is approximately $Q = 0$, the lines tangent to the circle labelled Cs serve to delineate the range of laboratory angles expected for the Cs product. For details, see S. Stolte, A. E. Proctor, and R. B. Bernstein, *J. Chem. Phys.* **65**, 4990 (1976).

matic aspects; a typical Newton diagram, that for the CsF + K system, is shown in Fig. 7.11. Both the Cs product and the CsF (elastically and inelastically scattered) are kinematically confined to a rather small angular cone in the laboratory system. The cone angle decreases with increasing E_{tr}.

Figure 7.12 shows representative laboratory angular distributions of scattered flux for each of the reactions 7.16 at low E_{tr}. The solid curves through the data points are calculated fits based on theoretically appropriate c.m. functions. From such data, it was possible to deduce the total cross sections for complex formation and the integral reactive cross sections; results are shown in the form of a log-log plot in Fig. 7.13.

More direct (and thus more accurate) ratio measurements yielded the so-called reactive branching fractions,

$$F_R = \frac{\sigma_R}{\sigma_R + \sigma_{NR}} = \frac{\sigma_R}{\sigma_C} \tag{7.17}$$

where σ_R is the integral cross section for reaction (i.e., to form KF), σ_{NR} that for the nonreactive scattering of the MF from the decay of the MFM′ complex, and $\sigma_C \equiv \sigma_R + \sigma_{NR}$ the cross section for complex formation (i.e., the capture cross section). These results are shown in Fig. 7.14. To be noted is the increase in reactivity with increasing E_{tr} for the endoergic reaction 7.16a, and the decrease for the exoergic 7.16b. This behaviour is qualitatively in accord with prior expectation, based on considerations applicable to a long-lived complex or a statistical model.

Figure 7.15 displays the laboratory angular distribution of the reactive and nonreactive scattering for the CsF + K reaction (Eqn. 7.16a), comparing the

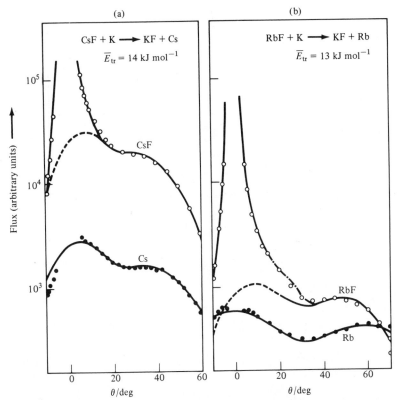

FIG. 7.12. Semilog plots of the laboratory angular distributions of scattering for the indicated MF + K reactions. *Open circles:* scattered flux of MF; *solid circles:* of product M. *Solid curves:* calculated fits using best c.m. function, transforming from c.m. to laboratory system. *Dashed curves:* calculated MF flux due to decay of MFK complexes. The large peak near 0° is due to quasielastic scattering of MF, well fit by theory. Adapted from S. Stolte et al., reference of Fig. 7.11; details therein.

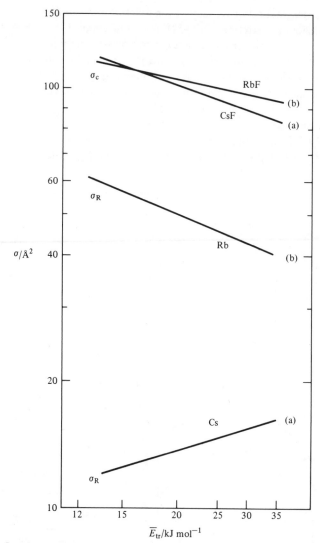

FIG. 7.13. Log-log plot of cross sections for reactions of CsF,RbF + K versus collision energy. Least squares lines have been drawn to represent experimental data. *Upper:* σ_C, cross section for capture (complex formation) for (a) CsF + K and (b) RbF + K. *Lower:* σ_R, cross section for reaction, i.e., (a) Cs formation and (b) Rb formation. Adapted from S. Stolte et al., reference of Fig. 7.11; details therein.

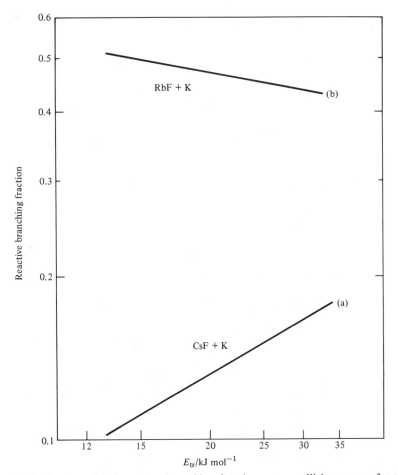

Fig. 7.14. Log-log plot of reactive branching fraction versus collision energy for the indicated reactions (a) and (b). Adapted from S. Stolte et al., reference of Fig. 7.11; details therein.

low-J CsF beam results with the thermal (Boltzmann) beam data. A substantial decline in the general level of reactivity is noted for the rotationally cold reagent. Analysis of the experimental data made it possible to evaluate the influence of the rotational energy difference ($\approx$ 9 kJ mol^{-1}) upon the total reaction cross section. This 'rotational energy effect' was expressed in terms of the reactive branching fractions (as defined by Eqn. 7.17), namely, the ratio F_R(low-J)/F_R(thermal). The results of such experiments carried out at different translational energies, for both reactions 7.16, are displayed in Fig. 7.16.

At the lowest translational energy, the effect is most pronounced: the reac-

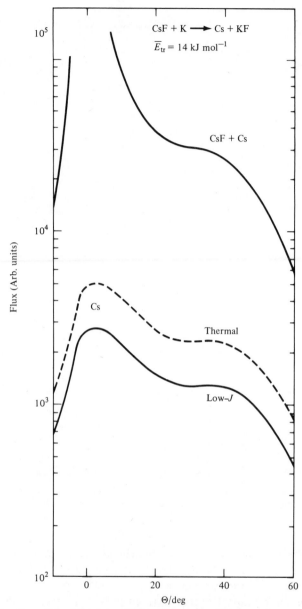

FIG. 7.15. Semilog plot of the laboratory angular distributions of scattering for the CsF + K reaction (smooth curves to represent the experimental data), using rotationally cold (low-J) CsF molecules. *Upper curve:* total flux of CsF + Cs. *Lower solid curve:* flux of Cs product from low-J CsF reagent. *Dashed curve:* Cs flux from thermal CsF reagent, for which $\overline{E}_{rot} \approx 10$ kJ mol^{-1}. Adapted from S. Stolte, A. E. Proctor, W. M. Pope, and R. B. Bernstein, *J. Chem. Phys.* **66**, 3468 (1977); details therein.

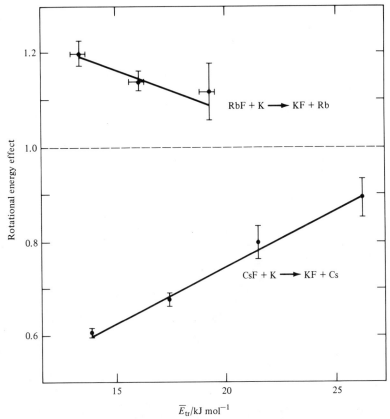

FIG. 7.16. Rotational energy effect upon reactive branching fraction versus collision energy for the two indicated reactions. Approximately 10 kJ mol^{-1} of rotational energy is removed from the reagent molecule via the low-J selection technique, resulting in the effects shown. Adapted from L. Zandee and R. B. Bernstein, *J. Chem. Phys.* **68**, 3760 (1978) and the results of S. Stolte et al., reference of Fig. 7.15; details therein.

tive branching fraction for the CsF reaction with rotationally cold CsF is only 60 per cent of its value for the thermal CsF; for RbF it is 120 per cent!

The explanation for these effects is rather simple. By combining the data on the translational energy dependence of the rotationally 'hot' and 'cold' beams, shown in Fig. 7.17 for Reaction 7.16a, it was shown that at constant *total* energy the effect of a change in reagent's rotational energy is comparable to that of a similar change in relative translational energy. Thus, to a first approximation, it is only the total energy ($E_{tr} + E_{rot} + E_{vib}$) of the colliding partners that governs the decay of the MFM′ complex. The rotational effect per se is more subtle and difficult to demonstrate unequivocally.

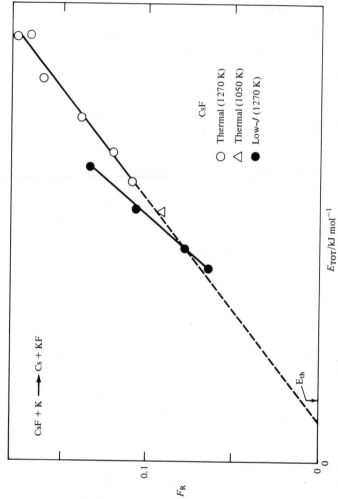

FIG. 7.17. Reactive branching fraction F_R versus total energy E_{tot} for CsF + K reaction, combining data from experiments with thermal and low-J CsF beams. Dashed line is extrapolation of least squares line through thermal data; its intercept is close to the thermodynamic threshold energy $E_{th} = \Delta E_0$. Adapted from S. Stolte et al., reference of Fig. 7.15.

7.6. Further vibrational enhancement studies via molecular beam and laser techniques

Next we turn to high-energy beam experiments on endoergic reactions and consider the dependence of the cross section for ion-pair formation upon the internal energy of the target molecule, as studied by J. Los et al. The reactions were of fast K atom beams with Br_2, CH_3I, and CH_3Br:

$$K + Br_2 \rightarrow K^+ + Br_2^- \tag{7.18a}$$

$$K + CH_3X \rightarrow K^+ + CH_3X^- \tag{7.18b}$$

The ionization cross sections were measured as a function of the temperature of the halogen compound, over a wide range of E_{tr}. At a given E_{tr} there appeared to be a nearly linear, positive temperature dependence.

The data were analyzed assuming that only vibration is responsible for the temperature dependence (i.e., ignoring the role of rotational energy). By the Hirschfelder–Eliason inverse Laplace transform technique, the specific vibrational state cross sections for reaction could be evaluated as a function of E_{tr}.

The results, plotted in Fig. 7.18, show that the vibrational effect is strongest near threshold and decreases with increasing E_{tr}. Internal energy appears to be much more efficient than translational energy in promoting ion-pair formation near threshold. Less than 1 eV of internal excitation produces more than an order of magnitude increase in cross section, for translational energies in the neighborhood of 5 eV. The strong vibrational effect has been interpreted in terms of 'prestretching' of the halogen bond with a resulting increase in the so-called curve-crossing radius R_c (and thus in the ionization cross section σ, given approximately by πR_c^2). See Chapter 8 for discussion.

Now we return to the more direct state-selective experiments which take advantage of laser-excited beams. An example is the vibrational enhancement study by R. N. Zare et al. of the reaction

$$HF(v) + Sr \rightarrow SrF + H \tag{7.19}$$

using laser-induced fluorescence (LIF) to detect the SrF product state distribution. Figure 7.19 shows the LIF spectrum of the ground electronic state SrF produced in Reaction 7.19 as a beam of Sr atoms passes through a target gas of HF. For the ground-state reagent $HF(v = 0)$, the reaction is endoergic by 27 kJ mol^{-1}, whereas for $HF(v = 1)$ it becomes 'exo' by 20 kJ mol^{-1}. The raw experimental data, even without analysis, demonstrate that the vibrational excitation energy of the HF has served to overcome the endoergicity (plus any activation barrier) of the reaction. The relative cross sections for the formation of the various vibrational states of the SrF product molecule contain valuable dynamic information.

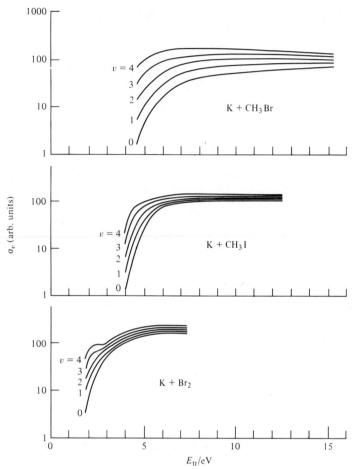

FIG. 7.18. Vibrational energy dependence of the cross section for collisional ionization as a function of the collision energy for the indicated reactions, derived from temperature dependence at given $\overline{E}_{tr}$. Adapted from A. M. Moutinho, A. W. Kleyn, and J. Los, *Chem. Phys. Lett.* **61**, 249 (1979); details therein.

A still more detailed laser excitation experiment on the same reaction has been carried out, but here two additional variables were probed: the rotational effect, i.e.,

$$HF(v = 1, J) + Sr \rightarrow H + SrF(v', J') \tag{7.20}$$

and the orientation effect. Polarized irradiation from a line-tunable HF laser was used to excite HF gas at a low pressure to a specified J state of the reactive $v = 1$ state; the excited HF served as a target for a Sr beam. The data provided an important piece of steric information, namely, that the reaction shows

$$Sr + HF(v) \longrightarrow SrF + H$$

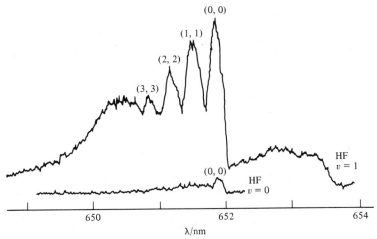

FIG. 7.19. Laser-induced fluorescence spectrum of the ground electronic state SrF formed in the endoergic Sr + HF reaction ($\Delta E_0 = 27$ kJ mol^{-1}) for HF($v = 0$) versus that for HF($v = 1$), exoergic by 20 kJ mol^{-1}. Adapted from Z. Karny and R. N. Zare, *J. Chem. Phys.* **68**, 3360 (1978); details therein.

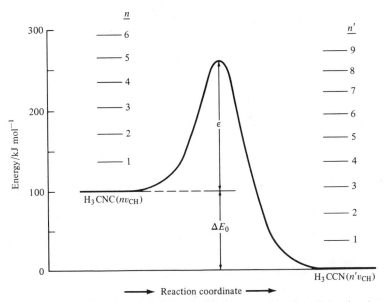

FIG. 7.20. Energetics of the methyl isocyanide isomerization reaction, showing the energy levels of the CH overtone-excited reagent and product molecules and the barrier to isomerization, $\epsilon_0 \approx 160$ kJ mol^{-1}. Adapted from K. V. Reddy and M. J. Berry, *Chem. Phys. Lett.* **52**, 111 (1977); details therein.

a preference for 'broadside attack.' The orientation effect on reactivity is becoming an important area of current research in molecular beam chemistry.

It was also found that there was a strong positive influence upon the cross section for populating higher *vibrational* levels in the SrF of higher *rotational* states of the HF(v = 1, J). The effect was stronger than anticipated purely on energetic grounds and was therefore attributed to the dynamic role of reagent rotation.

7.7. Vibrational enhancement of unimolecular reactions

All of the attention thus far has been on elementary bimolecular processes. However, there have been some very significant state-to-state studies of unimolecular reactions, carried out in the bulk gas phase using laser excitation. An example of such an investigation involving a state-selected unimolecular

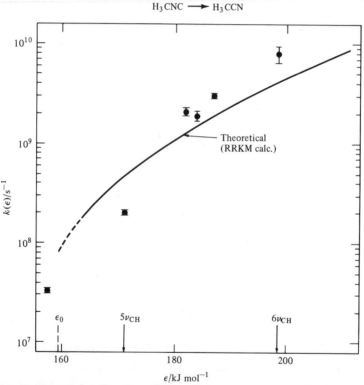

FIG. 7.21. Unimolecular rate constant for the isomerization of vibrationally state-selected methyl isocyanide. *Points:* experimental; *solid curve:* RRKM calculation. Adapted from reference of Fig. 7.20; details therein.

reagent is that of K. V. Reddy and M. J. Berry. The study involves the classical isomerization reaction of methyl isocyanide:

$$CH_3NC \rightarrow CH_3NC^{\dagger} \rightarrow CH_3CN \qquad (7.21)$$

where the first excitation step was accomplished by photoactivation, using intracavity cw dye laser irradiation (cf. Chapter 3). Successive overtones of the C-H bond stretch were excited, up to $6\nu_{CH}$ at $\lambda 621$ nm.

Figure 7.20 is a schematic diagram of the energetics of the system. The isomerization threshold is known from thermal isomerization kinetics to be $\epsilon_0 \approx 160$ kJ mol^{-1}. The energy levels shown are experimental values for the C-H stretching mode (overtones) in both $CH_3NC(n\nu_{CH})$ and $CH_3CN(n'\nu_{CH})$. Irradiation wavelengths were chosen in accordance with the measured absorption spectra. Seven irradiation wavelengths were used, e.g., $\lambda 793.7$ nm ($4\nu_{CH} + \nu_3$), 150.6 kJ mol^{-1}; $\lambda 726.6$ ($5\nu_{CH}$), 164.6 kJ mol^{-1}; etc., to $\lambda 621.4$ ($6\nu_{CH}$), 192 kJ mol^{-1}. Thus several total reagent energies (say ϵ) well above threshold (ϵ_0) were accessed by the overtone excitations.

A detailed analysis of the kinetic data on the isomerization made it possible to deduce the unimolecular rate coefficient, $k(\epsilon)$. Results are reproduced in Fig. 7.21. Also shown is a curve based on calculations using the RRKM theory. This experiment is important because the powerful laser excitation technique developed offers the possibility of a host of future studies of state-selected unimolecular reactions. Similar experiments on the selected-state isomerization of allyl isocyanide have shown small but possibly real differences in reactivity depending upon the particular type of C-H bond excited. There is some hope of demonstrating (and exploiting) bond-specific chemistry of polyatomic gases in the bulk via such tunable laser reagent excitation techniques.

In conclusion, it is clear that laser–molecular beam techniques have great potential as tools for the elucidation of the role of excited vibrotational states in chemical reactivity.

Review literature

ROB72, FOR73, FLY74, LEV74, POL74, BAE75, BER75x, BIR75, TRO75, TRU76, BER77, BRO77, BYL77, FAR77, GEI77, MOO77, POL77, TRU77, BAU78, RED78, BER79, FAR79, GLO79, KOM79a, LOS79, MOO79, RED79, ZAR79, CRO80, KNE80, LET80, PHY80, SMI80, ZAR80, EYA81, HOU81, JOR81, MCG81, SCH81

8 Electronically nonadiabatic collisions

8.1. Nonadiabatic collisions

Up to this point we have (tacitly) assumed the existence of a single potential energy hypersurface upon which the system 'evolves' from reagent(s) to product(s). The rolling of balls ('mass-points') on plaster surfaces to simulate classical trajectories is familiar to all students of chemical kinetics, and its success is a tribute to the insight of H. Eyring, M. Polanyi, and J. O. Hirschfelder, dating back to the late 1920's. Adiabatic potentials for diatomic molecules obtained with the use of the Born–Oppenheimer approximation form the basis for discussion for most of our diatomic spectroscopy, and conceptual extensions to adiabatic surfaces for stable polyatomic molecules and finally reacting systems have been quite natural and very useful for interpreting many if not most spectroscopic and dynamic properties.

Yet physicists studying high-energy collisions of electrons, ions, and atoms (and simple molecules) have, for five decades (since the papers of L. Landau and C. Zener) interpreted their results using diabatic potential curves (which can lead to 'curve-crossing') and, more generally, diabatic surfaces (which may intersect). As the collision energy increases (and thus the internuclear velocities become comparable to those of the valence electrons), the Born–Oppenheimer (B-O) separation approximation (electronic versus nuclear motion) becomes less valid, the B-O correction becomes much larger, and a diabatic treatment of the collision becomes preferable to an adiabatic one.

As chemists, we are mainly interested in a subclass of nonadiabatic collisions, namely, those involving atoms and molecules in which the electron distributions are 'reorganized' in some gross and easily detectable way in the course of the encounter.

Table 8-1 lists some common types of nonadiabatic collision processes which are of 'chemical' significance.

8.2. The two-state picture and the Landau–Zener approximation

The simplest case to discuss is that of atom-atom scattering in the so-called two-state picture. Shown in Fig. 8.1 are the divers potential curves needed to describe the two-state 'curve-crossing' problem. In the figure, $H_{11}(R)$ and $H_{22}(R)$ are the diabatic potentials for states 1 and 2, respectively, crossing at R_x, and $H_{12}(R)$ is the matrix element of the Hamiltonian that couples the two states, leading to a splitting between the adiabatic potentials $E_1(R)$ and $E_2(R)$.

TABLE 8-1
Some nonadiabatic collision processes

Process[a]	Identification[b]
Excitation and energy transfer	
A + BC ⌐→ A* + BC	T → E
└→ A + BC*	T → E
A* + BC ⌐→ A + BC*	E → E'
→ A + BC†	E → V'
└→ A + BC	E → T
A + BC* ⌐→ A* + BC	E → E'
→ A + BC†	E → V
└→ A + BC	E → T
Ionization, electron, and atom transfer	
⌐→ A⁺ + BC⁻	Collisional ionization
A + BC ⎤ → A⁺ + BC + e⁻	Penning ionization
A* + BC ⎬→ A + B⁺ + C⁻	Ion-pair formation
A + BC* ⎦ → AB⁺ + C⁻	Associative ionization
└→ AB⁺ + C + e⁻	Associative Penning ionization
(etc.)	

[a]The asterisk denotes electronic excitation; the dagger denotes vibrational (and rotational) excitation.
[b]The prime denotes transfer to partner.

The adiabats are given explicity in terms of the diabats by the relations:

$$E_{1,2} = 1/2\{H_{11} + H_{12} \pm [(H_{22} - H_{11})^2 + 4 H_{12}^2]^{1/2}\} \qquad (8.1)$$

where all energies are R dependent. The adiabats differ from the diabats only in the region of R near R_x, where $|H_{22} - H_{11}| \ll H_{12}$. At R_x, $H_{11}(R_x) = H_{22}(R_x) \equiv E(R_x)$ and thus

$$E_2(R_x) = E(R_x) + H_{12}(R_x) \qquad (8.2a)$$

$$E_1(R_x) = E(R_x) - H_{12}(R_x) \qquad (8.2b)$$

and so the splitting between the adiabats is

$$\Delta E(R_x) \equiv E_2(R_x) - E_1(R_x) = 2 H_{12}(R_x) \qquad (8.3)$$

The Landau–Zener approximation yields a simple expression for the probability of the system making a transition from diabat H_{11} to H_{22} at the crossing point R_x for a given value, say l, of the orbital angular momentum quantum number of colliding pair of atoms:

$$P_l \approx \exp(-K/v_l) \qquad (8.4)$$

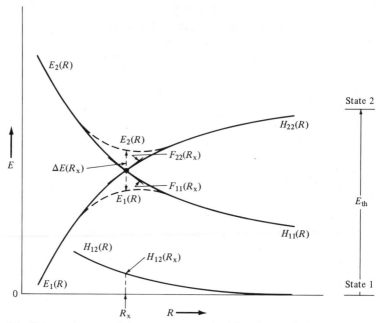

FIG. 8.1. Schematic potential curves appropriate for the analysis of the two-state atom-atom scattering problem. Here R_x is the crossing point where the two diabats $H_{11}(R)$ and $H_{22}(R)$, i.e., the potentials at large R, (would) intersect in the absence of any interaction. However, because of the finite interaction matrix element H_{12}, these curves 'repel one another' and split into the two adiabats $E_1(R)$ and $E_2(R)$. The energy separation between the adiabats is a minimum at R_x, with $\Delta E(R_x) = 2 H_{12}(R_x)$. See text for further details.

where v_l is the magnitude of the relative velocity calculated for the lth effective potential at R_x. Here, the positive constant K (in units of velocity) is a function of the potentials at R_x, as follows:

$$K = \frac{2\pi}{\hbar}\, [\, H_{12}(R_x)]^2/[F_{11}(R_x) - F_{22}(R_x)] = \frac{\pi^2}{h}\frac{[\Delta E(R_x)]^2}{[\Delta F(R_x)]} \qquad (8.5)$$

where $F_{11}(R_x)$ and $F_{22}(R_x)$ are the diabat forces (i.e., slopes of the potentials, $F = -dH/dR$ at R_x) and $\Delta E(R_x)$ the splitting between adiabats at R_x. Equation 8.4 implies that as v_l is increased from $0 \to \infty$, P_l increases rapidly from 0 to 1.

The *overall* probability of the system (of given l) jumping from state 1 to state 2 in the course of a complete traverse of A past B is given by

$$P_{1\to 2}^{(l)} = 2\, P_l(1 - P_l) \qquad (8.6)$$

Equation 8.6 follows from a consideration of the joint probabilities for the 'approach' and for the 'departure' passage across R_x. (Note that $P_{1\to2}$ achieves its maximum value at $P_l = \frac{1}{2}$.)

Summing over all partial waves yields the cross section for making the excitation transition from asymptotic state 1 to state 2:

$$\sigma_{1\to2} = \frac{\pi}{k_i^2} \sum_{l=0}^{\infty} (2l + 1) P_{1\to2}^{(l)} \to \int_{b=0}^{\infty} db\, 2\pi b P_{1\to2}(b) \qquad (8.7)$$

where $k = \mu v/\hbar$ and $b = (l + \frac{1}{2})/k$, as usual. The integral has been worked out, yielding a general expression for the cross section (derived by D. R. Bates and T. J. Boyd):

$$\sigma_{1\to2} = \pi R_x^2 4\, G(v/K) \left[\frac{E - E(R_x)}{E_{tr}} \right] \qquad (8.8)$$

for $v \geq v_{th}$, i.e., $E_{tr} \geq E_{th}$, where E is the total energy and $E_{tr} = (\frac{1}{2})\mu v^2$ is the relative translational energy. For the excitation process, $E = E_{tr}$ and the term in brackets becomes simply $1 - E(R_x)/E_{tr}$. The function $G(v/K)$, a function of the quantity K of Eqn. 8.5 and the incident relative velocity v, is shown in Fig. 8.2: it is zero for $v \leq v_{th}$ (where the threshold velocity is given

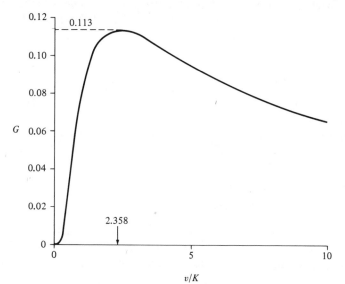

FIG. 8.2. Plot of the function G (of Eqn. 8.8) versus the reduced speed v/K. This functionality, appropriate for the Landau–Zener approximation, was first derived by D. R. Bates and T. J. Boyd, *Proc. Phys. Soc. London* **A69**, 910 (1956); see also D. R. Bates, *Comm. At. Mol. Phys.* **2**, 107 (1970). Adapted from BAE75.

directly in terms of the threshold energy: $v_{th} = (2E_{th}/\mu)^{1/2}$, with E_{th} the asymptotic separation of states 1 and 2; cf. Fig. 8.1) and rises to a maximum of 0.113 at $v = 2.358 \, K$. Of course, for $v \leq v_{th}$, $\sigma_{1 \to 2} = 0$.

The de-excitation cross section $\sigma_{2 \to 1}$ is finite for all v (i.e., there is no energetic threshold). It is given by

$$\sigma_{2 \to 1} = \pi R_x^2 4 \, G(v/K) \left[\frac{E - E(R_x)}{E'_{tr}} \right] \tag{8.9}$$

where $E'_{tr} = E - E_{th} = \tfrac{1}{2}\mu v'^2$ is the relative translational energy for the de-excitation collision. At the same total energy, the cross sections satisfy the fundamental microreversibility condition:

$$E_{tr}\sigma_{1 \to 2}(E) = E'_{tr}\sigma_{2 \to 1}(E) \tag{8.10}$$

Whereas the excitation cross section increases abruptly from zero to a finite value at threshold, the de-excitation cross section declines sharply with translational energy from its very large value near $E'_{tr} = 0$ to a value approaching the excitation cross section at very high energies (cf. Eqn. 8.10 in the limit $E \gg E_{th}$, where $(E'_{tr} - E_{tr})/E \to 0$).

Exact quantum mechanical calculations using the so-called coupled-channels approach have been carried out for a number of model systems by R. D. Levine, B. R. Johnson, M. B. Faist, and others. Results have served to confirm the validity of the LZ treatment as a quite good approximation.

8.3. Atom-atom scattering and collisional ionization

Now consider the scattering of atom A by atom B, or ion A^+ by ion B^-, as discussed in MRD. Figure 8.3 shows the two relevant diabatic potential curves, here denoted $V_1(R)$ and $V_2(R)$. The asymptotic separation ΔE_0 of the endoergicity of the collisional ionization (charge transfer) reaction $A + B \to A^+ + B^-$ is given by the difference between the ionization potential of $A(IP_A)$ and the electron affinity of B (EA_B), i.e.,

$$\Delta E_0 = IP_A - EA_B \tag{8.11}$$

The so-called lower diabat $V_1(R)$ is usually a repulsive curve (except at large separations R, not shown, where there is a shallow van der Waals potential well). The upper diabat is taken to be a coulombic, ion-ion, attractive curve,

$$V_2(R) = \Delta E_0 - e^2/R \tag{8.12}$$

which crosses the zero at a separation R_0

$$R_0/nm = e^2/\Delta E_0 = 1.435/\Delta E_0 \,(eV) \tag{8.13}$$

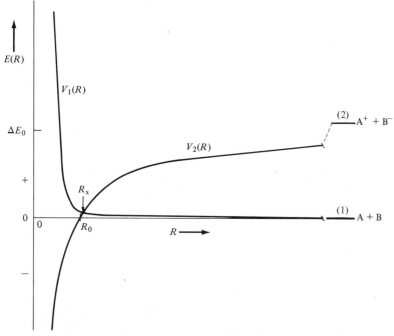

FIG. 8.3. Schematic potential curves appropriate for the LZ description of the charge transfer (collisional ionization) process $A + B \rightarrow A^+ + B^-$. The lower diabat $V_1(R)$ is a covalent curve with a shallow van der Waals minimum at large R. The upper diabat $V_2(R)$ is taken as a coulombic curve (Eqn. 8.12). The 'avoided crossing' is at $R_x \approx R_0$ (Eqn. 8.13). See text for details.

For the case shown, where $V_1(R)$ is nearly flat at the zero of potential, the separation at which the diabatic curves cross, R_x, is only slightly larger than the R_0 of Eqn. 8.13. For the limiting case, in which, at R_x, the covalent potential is flat at zero and the ionic potential is inverse first power (coulombic), the slope difference in Eqn. 8.5 can be written down immediately and so K becomes simply a function of $H_{12}(R_x)$ and ΔE_0:

$$K \approx 4\pi^2 R_x^2 [H_{12}(R_x)]^2 / e^2 h = 4\pi^2 (e^2/h)[H_{12}(R_x)/\Delta E_0]^2 \quad (8.14)$$

in units of velocity as required by Eqn. 8.4. Note that in the limit of small splittings, i.e., $H_{12}(R_x)/\Delta E_0 \approx 0$, $K \rightarrow 0$ and thus v/K becomes large, so that v/K will lie far to the right of its maximum (Fig. 8.2) and thus $G(v/K)$ and $\sigma_{1\rightarrow2}$ (from Eqn. 8.8) will be very small.

As alluded to earlier, theory dictates that two such diabatic potentials (corresponding to states of the same symmetry) which interact with a matrix element $H_{12}(R)$ have an 'avoided crossing,' which yields the splitting $\Delta E(R_x)$

between the two adiabatic curves. Given the splitting and the slopes of the diabats at R_x, one can calculate the cross section using the LZ approximation, Eqn. 8.8 (using Eqn. 8.5 for the quantity K). However, it is not yet possible to predict the potentials *ab initio* with sufficient accuracy to obtain a useful estimate of K, and so it is usual to evaluate K empirically by fitting the experimental cross section versus velocity data to the LZ form, using a log-log plot of $\sigma_{1\to 2}$ versus v.

Figure 8.4 is such an example, in which the fit of the data to the LZ functional form makes possible the determination of K from the abscissa (and, if absolute cross sections are known, from the ordinate fit). The experiments were of the atomic beam, collisional ionization type, done at the FOM Laboratory in Amsterdam using the apparatus sketched in Fig. 8.5.

Figure 8.6 shows diabatic potentials for the Na + I system, where R_x is approximately given by the R_0 of Eqn. 8.13. The value of $H_{12}(R_x \approx 0.7 \text{ nm})$ turns out to be about 0.1 eV for this system, and so the splitting of the adiabats (≈ 0.2 eV) could hardly be perceptible on the scale of the figure. For the case of K + I, $\Delta E_0 = 4.34 - 3.06 = 1.28$ eV, so that $R_x \approx 1.13$ nm (via Eqn.

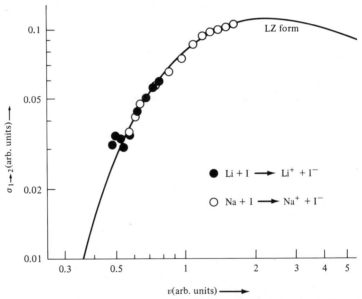

FIG. 8.4. Fit of experimental charge transfer cross sections (for the indicated reactions) to the LZ cross section function. The ordinate (log scale) is the cross section in units of $4\pi R_x^2$; the abscissa (log scale) is the reduced speed v/K. Thus the fit for a given reaction enables the value of K and thus of $H_{12}(R_x)$ to be determined. Adapted from A. M. Moutinho, J. A. Aten, and J. Los, *Physica* **53**, 471 (1971); see also BAE75.

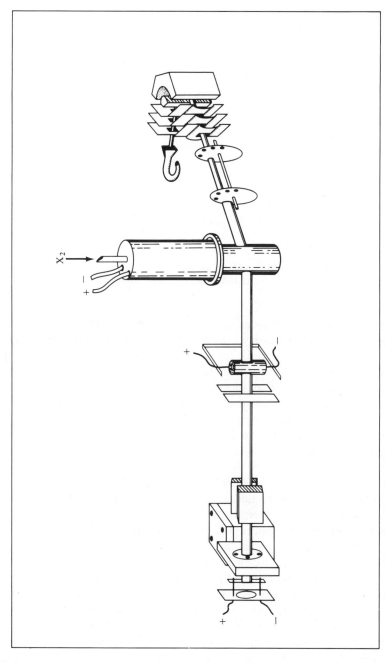

FIG. 8.5. Schematic diagram of the FOM Laboratory differential scattering apparatus for measurement of ion-pair formation from collisions of fast alkali atoms with halogens. Adapted from J. A. Aten and J. Los, *Chem. Phys.* **25**, 47 (1977); see also LOS79; details therein.

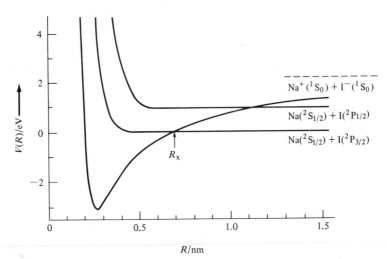

FIG. 8.6. Potential curves for the $^1\Sigma^+$ states of the Na $+$ I system. The lower covalent curve intersects the ionic curves at approximately 0.7 nm. Adapted from A. P. Baede, BAE75.

8.13). From the fit of the collisional ionization cross sections to the LZ energy dependence, one obtains a splitting $\Delta E(R_x)$ of only 2.5 meV. Generally speaking, $H_{12}(R_x)$ values are very much smaller for large R_x crossings.

To be noted on Fig. 8.6 is another diabat, for an excited state of the Na atom. The implications of the additional avoided crossing (at larger R) will be discussed later. It is already clear that even for such simple atom-atom systems multiple curve crossings must be taken into consideration. Figure 8.7 represents the simplest situation for the hypothetical A $+$ B system, in which one excited state for A and one for B are included (in addition to a stylized repulsive curve for the ground-state A–B interaction and a coulombic A^+B^- potential). Obviously, a two-state theory will be inadequate for dealing accurately with realistic multistate interactions. However, the saving feature is that the avoided crossings at large R_x usually have associated with them small interaction matrix elements and splittings $\Delta E(R_x)$. The problem of multiple crossings is the subject of current research.

Returning to the simple two-state collisional ionization problem, it is of interest to consider angular distributions as well as total ionization cross sections, as done by J. Los, A. W. Kleyn, G. A. Delvigne, A. P. Baede, and coworkers. In this connection, Fig. 8.8 is illustrative. It shows that four distinguishable trajectories can result from a fast atom-atom collision at a given impact parameter. When the nuclei approach (along the covalent potential) and reach the separation R_x, there can be either 'covalent' or 'ionic' scattering; similarly on the retreat (separation) path. Clearly, such ionic and covalent tra-

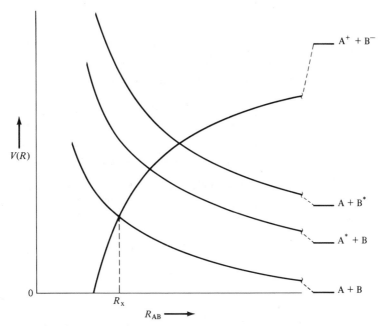

FIG. 8.7. Schematic potential curves (diabats) for a hypothetical A + B system, showing the various crossings with the ionic curve of the repulsive potentials for the ground-state atoms (A + B), for excited A with B and for excited B with A. See text.

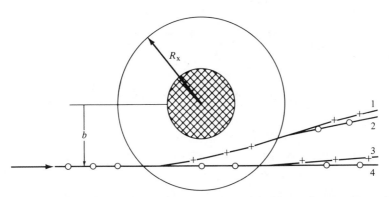

FIG. 8.8. Geometric interpretation of the four possible outcomes of a collision of a fast alkali atom with a fixed target halogen atom. The crossing point (separation) R_x (Fig. 8.1) fixes the radius of the outer circle (the inner circle corresponds to the repulsive core). The open symbol denotes neutral; the plus sign indicates a trajectory in which an ion pair has been formed. If an electron jump occurs at the first crossing *and* the system remains ionized at the second, the result for the ionic trajectory corresponds to wide-angle scattering (labelled 1). For trajectory 2, a second electron jump has occurred at the second crossing, resulting in wide-angle neutral scattering. Trajectories 3 and 4 have obvious implications. Adapted from J. Los and A. W. Kleyn, LOS79; details therein.

jectories imply interesting differential scattering behaviour, including interference phenomena, as shown in Fig. 8.9 for the Na + I reaction. It is beyond the scope of this chapter to deal with the angular distributions, but it turns out that they can be quite well accounted for with simple theory, making use of an extension of the LZ approximation due to E. C. Stueckelberg (which requires no further information than LZ itself!).

8.4. Curve-crossing spectroscopy and spin-orbit transitions: IBr

Next we shall look at a simple atom-atom system (IBr) in which the spin-orbit excitation energy of the Br atom gives rise to interesting curve-crossing spectroscopy and dynamics. Figure 8.10 is a schematic diagram of the low-lying

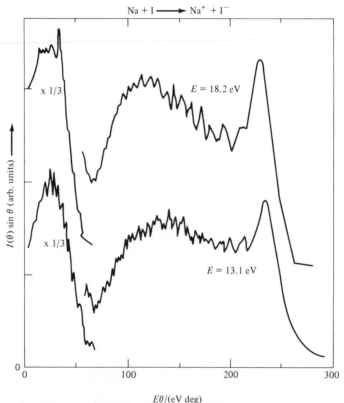

FIG. 8.9. Experimentally measured differential cross sections (converted to the c.m. system) for ion-pair formation from Na + I collisions at the two designated collision energies E, plotted versus the variable $E\theta$. The coarse (rainbow) structure as well as the fine (interference) structure in the data is in accord with calculations based upon an extended LZ-type theory. Adapted from G. A. Delvigne and J. Los, *Physica* **67**, 166 (1973); details therein.

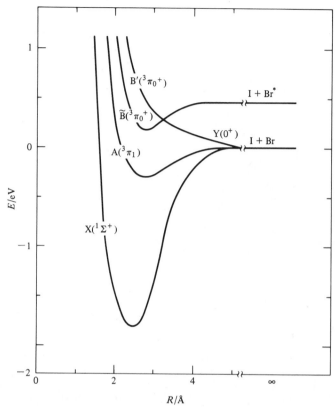

E/eV

$R/\text{Å}$

FIG. 8.10. Approximate diagram of some of the low-lying potential curves for IBr. The asterisk denotes the upper spin-orbit state Br $(^2P_{1/2})$. Adapted from M. B. Faist and R. B. Bernstein, *J. Chem. Phys.* **64**, 2971 (1976); details therein. See also M. S. Child and R. B. Bernstein, *J. Chem. Phys.* **59**, 5916 (1973).

potential curves for IBr. The avoided crossing of the diabats dissociating to I + Br* and I + Br is of interest. Figure 8.11 is a detailed drawing of the two adiabats and their 'precursor' diabats. From the known vibrational levels in the wells of both the lower and the upper adiabats and from the known asymptotic energy for I + Br*, the position and depth of the minimum in each adiabat and then the splitting $\Delta E(R_x)$ can be estimated. Thus one can construct the approximate potentials shown. LZ approximation calculations of the excitation and de-excitation cross sections, respectively,

$$\text{I} + \text{Br} \rightarrow \text{I} + \text{Br*} \quad \text{excitation} \qquad (8.15a)$$

$$\text{I} + \text{Br*} \rightarrow \text{I} + \text{Br} \quad \text{de-excitation} \qquad (8.15b)$$

are shown in Fig. 8.12a. They have been compared with 'exact' two-state calculations; the results are displayed in Fig. 8.12b. Except for energies just above

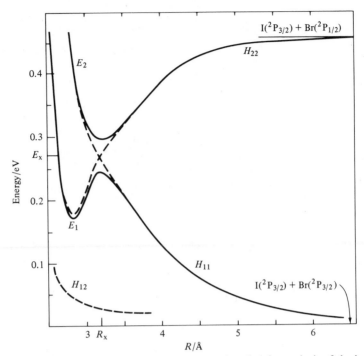

FIG. 8.11. Diagram of the potential curves appropriate for the analysis of the inelastic (spin-orbit transition) scattering problem $I + Br(^2P_{3/2}) \rightleftharpoons I + Br(^2P_{1/2})$. Solid lines denote the adiabats E_1 and E_2, dashed lines the diabats H_{22} and H_{11} and the interaction matrix element H_{12}. The R_x and E_x denote the coordinates of the (diabatic) crossing point shown. Adapted from the reference of Fig. 8.10; details therein.

threshold, the LZ approximation is entirely adequate. Similarly, inelastic scattering cross sections for many other processes can be described by the LZ curve-crossing approximation. M. S. Child, A. D. Bandrauk, and others have developed optimal semiclassical treatments of curve-crossing spectroscopy.

It is of interest to note that one can readily obtain simple expressions for the bimolecular rate constants for such curve-crossing, inelastic collisions. To see the systematics of the so-called LZ rate constant, it is best if one divides the field into two types of potential curve-crossings as portrayed schematically in Fig. 8.13.

The first type (the 'attractive' case) is the one discussed earlier at some length, characterised by a curve-crossing at an energy $E(R_x)$ below the threshold energy E_{th}. The second type (the 'repulsive' case) is one in which the curve-crossing occurs at an energy *above* the threshold (i.e., above the endoergicity ΔE of the process). In contrast to the previous case, this leads to an (activation) energy barrier, say $D \equiv E(R_x) - \Delta E_0$. In Fig. 8.13, attention is confined to the de-excitation process (in both cases). A reaction coordinate representation is presented, which shows the underlying reason for an important difference in the behaviour of the inelastic rate constants for the two cases.

It has been shown that the excitation and de-excitation rate coefficients, respectively, for the 'attractive' (a) case are of the form

$$k_{1\to2}(T) = A^{(a)}(T) \tag{8.16a}$$

$$k_{2\to1}(T) = A^{(a)}(T) \exp(-\Delta E_0/kT) \tag{8.16b}$$

and for the 'repulsive' (r) case,

$$k_{1\to2}(T) = A^{(r)}(T) \exp(-D/kT) \tag{8.17a}$$

$$k_{2\to1}(T) = A^{(r)}(T) \exp[-(D + \Delta E_0)/kT] \tag{8.17b}$$

where $A^{(a)}(T)$ and $A^{(r)}(T)$ are slowly varying, known functions of temperature T (proportional to $R_x^2 K$). In the high-T limit, these LZ pre-exponential factors become temperature independent and simple 'Arrhenius' equations result.

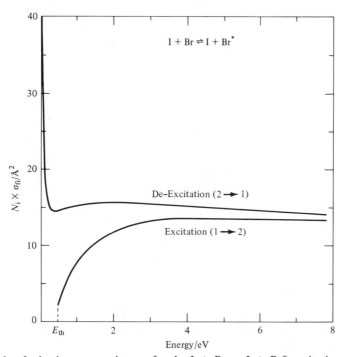

FIG. 8.12a. Inelastic cross sections σ_{fi} for the I + Br $\to$ I + Br* excitation and de-excitation processes. Plot of LZ-approximated cross sections versus collision energy (ordinate contains statistical factors, $N = 16,8$ for $i = 1,2$, respectively.

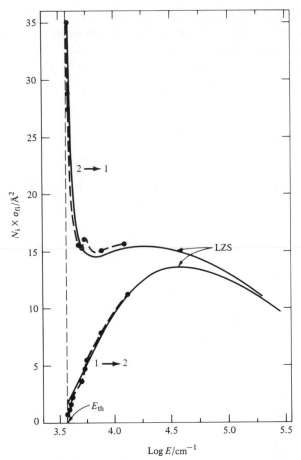

FIG. 8.12(b). Exact quantal-calculated cross sections (points) plotted versus log E, compared with LZ-approximated results (solid curves). Adapted from reference of Fig. 8.10; details therein.

8.5. Atom-molecule scattering and collisional ionization

Next we turn to nonadiabatic atom-diatom collisions. For simplicity we consider the interaction of alkali atoms with halogen molecules, e.g., the collisional ionization (charge transfer) process:

$$K + Br_2 \rightarrow K^+ + Br_2^- \tag{8.18}$$

As a first approximation, one can assume a constant internuclear separation of Br_2 and represent the system by two diabatic potential curves as sketched in Fig. 8.14 (see MRD).

Provided the relative translational energy exceeds the threshold $\Delta E_0 = IP(K) - EA(Br_2) = 4.34$ eV $- EA(Br_2)$, some ions will be formed. In fact, measurement of such threshold energies provides an excellent means of determining the electron affinities of the target molecules, as shown in Fig. 8.15 for the reaction shown in Eqn. 8.18 (and for Na and Li projectiles as well). (This procedure was first implemented by R. K. Helbing and E. W. Rothe and by J. Los, A. DeVries, and co-workers.) From the three values of ΔE_0, using Eqn. 8.11 with known ionization potentials for the three alkalis, A. P. Baede obtained three independent values of the electron affinity of Br_2, all within the range 2.55–2.60 eV. Figure 8.16 shows similar data for the heteronuclear halogen molecule IBr.

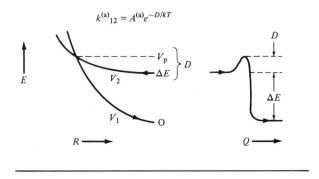

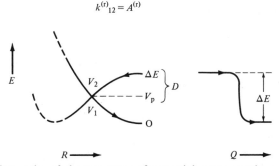

FIG. 8.13. Energetics of the two types of potential curve crossings. *Upper:* crossing point $[V_1(R_x) = V_2(R_x) = V_p]$ above the endoergicity; *lower:* $V_p < \Delta E$. Right-hand part of figure shows reaction coordinate (Q) representation; the upper one shows an 'activation barrier' $D \equiv V_p - \Delta E$ not found in the lower one. Adapted from M. B. Faist and R. B. Bernstein, *J. Chem. Phys.* **64**, 3924 (1976); details therein.

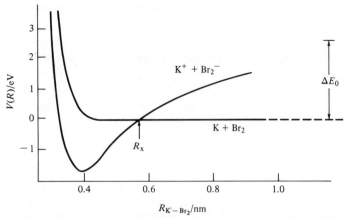

FIG. 8.14. Schematic drawing of potential curves to represent the harpoon reaction of K + Br_2, appropriate to describe both the subthreshold reaction to yield KBr + Br and the postthreshold reaction forming $K,^+ Br_2^-$ products. The lower curve corresponds to the covalent ($K-Br_2$) interaction at large separations, the upper one to the ionic ($K^+-Br_2^-$) potential. As R decreases to the crossing point R_x, the system can 'switch' from covalent to ionic and may follow the lower curve to form K^+Br^-, ejecting Br. The energy threshold $\Delta E_0 = IP(K) - EA(Br_2)$. Adapted from MRD, Section 4.1.4; details therein.

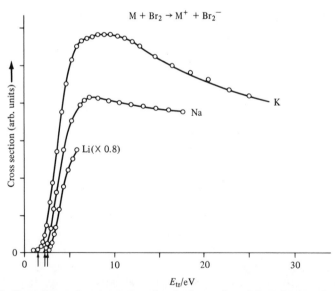

FIG. 8.15. Experimental cross sections for the formation of negative ions by collision of fast alkali atoms (M ≡ K, Na, Li) with Br_2 as a function of relative translational energy. Energy thresholds are indicated by the arrows. Adapted from A. P. Baede and J. Los, *Physica* **52**, 422 (1971); see also BAE75; details therein.

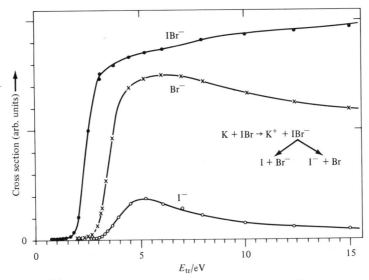

FIG. 8.16. Experimental cross sections for the formation of the indicated negative ions by collision of fast K by IBr, versus E_{tr}. Note the branching ratio (Br$^-$/I$^-$). Adapted from D. J. Auerbach, M. M. Hubers, A. P. Baede, and J. Los, *Chem. Phys.* **2**, 107 (1973); details therein.

Based on detailed considerations, it has been concluded that the electron affinities obtained from collisional ionization thresholds are true 'adiabatic' rather than 'vertical' affinities. Since these are real thermodynamic properties, one can use these quantities to calculate the binding energies of the diatomic anions. Relevant potential curves are portrayed schematically in Fig. 8.17.

Consider the thermochemical cycle as follows:

$$\underline{\Delta E}$$

$$\begin{array}{ll} AB = A + B & D_0(AB) \\ e^- + B = B^- & -EA(B) \\ \underline{AB^- = AB + e^-} & \underline{EA(AB)} \\ AB^- = A + B & D_0(AB^-) \end{array}$$

Thus

$$D_0(AB^-) = D_0(AB) + EA(AB) - EA(B) \qquad (8.19)$$

Since the dissociation energy of the neutral AB molecule is known, as is, of course, the electron affinity of atom B, the newly obtained electron affinity of AB yields the hitherto unknown dissociation energy of AB$^-$.

Table 8.2 summarizes results for the halogens.

Molecular electron affinities can also be obtained from endoergic charge transfer reaction thresholds, using negative atomic ion beams, e.g.,

$$X^- + AB \rightarrow X + AB^- \qquad (8.20)$$

as first accomplished by W. A. Chupka, J. Berkowitz, and others. The endoergicity of the reaction is simply

$$\Delta E_0 = EA(X) - EA(AB) \qquad (8.21)$$

Provided there is a negligible activation barrier for Reaction 8.20, the measured threshold energy will be equal to ΔE_0, and one can calculate $EA(AB)$

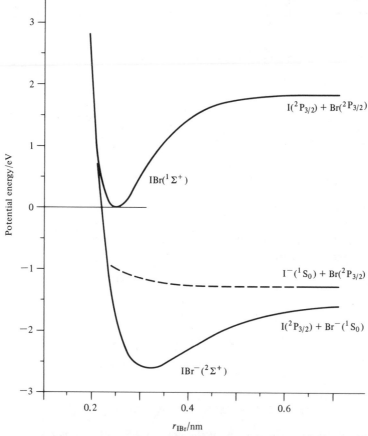

FIG. 8.17. Potential curves for IBr and IBr⁻, illustrating the problem of the definition of a molecular electron affinity. The asymptotic separation between the lowest curves is 0.30 eV. Adapted from LOS79, details therein; based on reference of Fig. 8.16 and M. M. Hubers, A. W. Kleyn, and J. Los, *Chem. Phys.* **17**, 303 (1976).

TABLE 8.2
Electron affinity of halogen
molecules and dissociation energy
of halogen molecule anions
(BAE75)

AB	$EA(AB)/eV$	$D_0(AB^-)/eV$
Cl_2	2.45	1.31
Br_2	2.55	1.15
I_2	2.55	1.02
ICl	2.41	0.95
IBr	2.55	1.05

directly. As in the collisional ionization experiments, a consistency check is available by comparison of the $EA(AB)$ values obtained from experiments with several different anions (e.g., I^-, Br^-, Cl^-, O^-) Results are quite self-consistent, and, moreover, the $EA(AB)$ values thus derived for the halogen molecules accord well with those from the alkali collision results (Table 8.2).

Similar studies on alkali-molecule charge transfer collisions yield electron affinities for various nonhalogen diatomics, e.g., O_2 ($EA \approx 0.5$ eV), NO (0.1 eV), and for several polyatomic molecules, e.g., CH_3I ($EA \approx 0.3$ eV), SF_6 (0.5 eV), and NO_2 (2.5 eV).

In addition to the determination of thresholds and thus energetics of the alkali charge transfer reactions, it has been possible to extract values of R_x and $H_{12}(R_x)$ from LZ fits to the cross section–velocity dependencies. Based on considerable theoretical and semiempirical effort, a 'reduced' semilogarithmic relationship has been established, correlating all results for alkali (M) – halogen molecule (XY) systems, as plotted in Fig. 8.18. The ordinate is a 'reduced coupling matrix element,' defined in terms of $H_{12}(R_x)$ by

$$H_{12}^* \equiv H_{12}(R_x)[IP(M) - EA(XY)]^{1/2} \qquad (8.22a)$$

The abscissa is a reduced crossing point:

$$R_x^* \equiv R_x[2\,IP(M)]^{1/2} \qquad (8.22b)$$

The line passing nicely through all the data is represented by

$$H_{12}^*/R_x^* = 1.73\exp(-0.875\,R_x^*) \qquad (8.23)$$

Thus the field of alkali charge transfer reactions is at a rather advanced level of systematization.

Another type of nonadiabatic process that has received attention by molec-

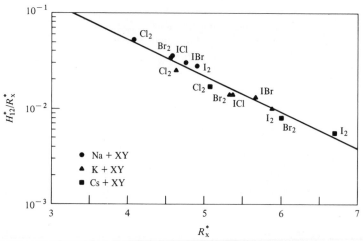

FIG. 8.18. Semilog plot of the 'reduced' LZ coupling matrix element H_{12}^* versus reduced crossing point R_x^*, here defined via Eqns. 8.22 and 8.23. See text. Adapted from LOS79. Theory of R. E. Olson, F. T. Smith, and E. Bauer, *Appl. Opt.* **10**, 1848 (1971); data of M. M. Hubers, A. W. Kleyn, and J. Los, *Chem. Phys.* **17**, 303 (1976); details therein.

ular beam methods is that of collision-induced ion-pair formation, e.g., the dissociation of alkali halides into ions:

$$Xe + CsCl \rightarrow Xe + Cs^+ + Cl^- \qquad (8.24)$$

first studied by Y. T. Lee, R. S. Berry et al. and by S. Wexler, E. K. Parks, and co-workers. The threshold energy (determined as usual by increasing the relative translational energy until the appearance of ions) is essentially the endoergicity ΔE_0, which is of course the dissociation energy of the salt into ion pairs, say $D_0(M^+X^-)$, related to $D_0(MX)$ by

$$D_0(M^+X^-) = D_0(MX) + IP(M) - EA(X) \qquad (8.25)$$

Experimental data for Reaction 8.24 are represented in Fig. 8.19. The dashed curves represent results of a deconvolution of the data from the velocity spread of the beams, i.e., the best estimate of the true cross sections. Shown also in the figure are results for a concomitant ion transfer or abstraction reaction (with a much smaller reaction cross section):

$$Xe + CsCl \rightarrow CsXe^+ + Cl^- \qquad (8.26)$$

8.6. The harpoon mechanism

Thus far, we have emphasized nonadiabatic processes leading to charge transfer, for example, Reaction 8.18 of $K + Br_2$ producing K^+ and Br_2^-. Referring back to Fig. 8.14, we see that for $E < \Delta E_0$ no ions can be formed. We know, however, that the reaction of K + Br$_2$ to produce neutral products

$$K + Br_2 \rightarrow KBr + Br \qquad (8.27)$$

is *exoergic:*

$$-\Delta E_0 = D_0(KBr) - D_0(Br_2) \approx 2.0 \text{ eV}$$

As discussed in Chapter 6, etc., these alkali-halogen reactions have no observable thresholds and large reaction cross sections, which decline slightly with increasing E_{tr}. The microscopic description of these reactions is via the so-called harpoon mechanism originally enunciated by M. Polanyi 50 years ago.

As the neutral particles (atom and molecule) approach along the nearly flat covalent curve (cf. Figs. 8.3, 8.14) at a critical separation (R_x), there is an effective transfer of an electron from the alkali to the halogen (the alkali tosses the harpoon into the halogen) to form an ion pair. Then the strong coulombic attractive force (due to the lower, ionic potential) accelerates the M$^+$ ion towards the X$_2^-$ ion (the harpoon is being pulled in!), and so reaction occurs. The alkali atom has thus used its valence electron as a harpoon to draw the halogen in close, allowing the transfer of X$^-$ to form the neutral alkali halide

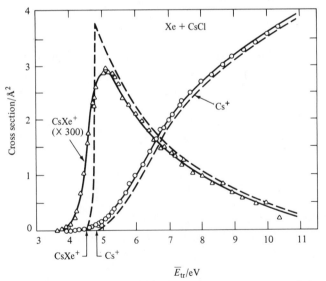

FIG. 8.19. Experimental measurements of absolute cross sections for Cs$^+$ and CsXe$^+$ formation from collisions of CsCl + Xe as a function of the average relative translational energy (points and solid curves). Dashed curves are deconvoluted results from which the two indicated thresholds have been deduced. Adapted from S. H. Sheen, G. Dimoplon, E. K. Parks, and S. Wexler, *J. Chem. Phys.* **68**, 4950 (1978).

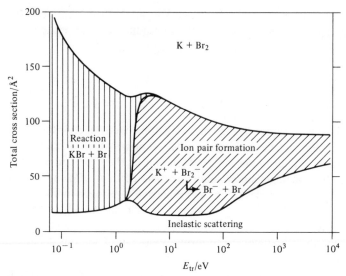

FIG. 8.20. Overall summary of experimentally determined cross sections for the various indicated nonelastic processes for the K + Br$_2$ system, as a function of the collision energy over a range of five orders of magnitude in E. Based on C. W. Evers, *Chem. Phys.* **30**, 27 (1978); adapted from LOS79; details therein.

salt MX, releasing a neutral halogen atom X. (See MRD for further discussion.)

Using the crudest of arguments, it can be seen that the harpoon reaction cross section is bound to be approximately

$$\sigma_R \approx \pi R_x^2 \qquad (8.28)$$

(assuming the reaction probability to be unity for impact parameter $b \le b_c$), and so $\sigma_R = \pi b_c^2$, where b_c is the critical value of b for which the turning point is R_x:

$$b_c^2 = R_x^2 \left[1 - \frac{V(R_x)}{E_{tr}} \right] \qquad (8.29)$$

For $V(R_x)/E_{tr} \approx 0$, $b_c \approx R_x$ and Eqn. 8.28 ensues.

Using Eqns. 8.13 and 8.28, the harpoon model yields reaction cross sections given by

$$\sigma_R = \pi \left[\frac{14.35}{\Delta E_0} \right]^2 \qquad (8.30)$$

where ΔE_0 is in eV and the units of σ_R are Å^2. For this alkali-halogen reactions with large values of R_x, cross sections range from 100 Å^2 (Li + I$_2$) to 200 Å^2

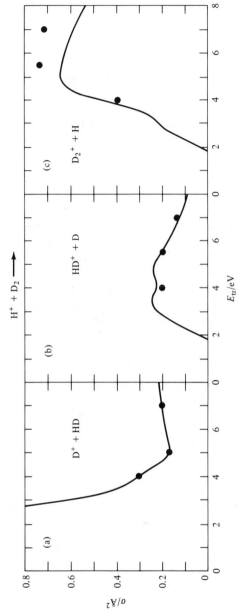

FIG. 8.21. Absolute cross sections as a function of collision energy for the indicated reaction products for $H^+ + D_2$ reactions yielding (a) $D^+ + HD$, (b) $HD^+ + D$, and (c) $D_2^+ + H$. *Points*: calculated by the trajectory surface-hopping method by J. C. Tully and R. K. Preston, *J. Chem. Phys.* **55**, 562 (1972); *solid curves*: experimentally determined results by G. Ochs and E. Teloy, *J. Chem. Phys.* **61**, 4930 (1974).

(Cs + Br$_2$), as reviewed by P. Davidovits (DAV79). The differential scattering cross sections are of course forward-peaked as a consequence of the long-range crossings, i.e., the harpooning taking place at large (0.5–0.15 nm) separations of the neutral reagents. Large b yields low angle scattering and vibrationally excited product, as discussed in Chapter 6.

Figure 8.20 summarizes, in a global way, the nonelastic scattering behaviour of the K + Br$_2$ system, studied over an unusually wide energy range (10^{-1}– 10^4 eV!). The reaction to form neutral products gives way to the ion-pair reaction above the collisional ionization threshold; the inelastic scattering contribution grows at higher E_{tr}.

8.7. Intersecting potential surfaces and 'trajectory surface hopping'

Finally, we must turn our attention to the more detailed 'polyatomic' aspects of nonadiabatic reactions. As pointed out in MRD, one can go only so far with a two-body approach to a reaction involving three or more atoms. Exact *ab initio,* quantum mechanical solutions are of course not yet available, but for a few simple systems very good approximate theoretical treatments have been successfully carried out. One of these is for reactions involving the simple stable triatomic ion H$_3^+$ and its isotopic variants. The HD$_2^+$ ion can be considered the reaction intermediate for a number of concomitant H$^+$ + D$_2$ reactions:

$$
\text{H}^+ + \text{D}_2 \rightarrow [\text{HD}_2^+] \begin{array}{l} \overset{(a)}{\longrightarrow} \ \text{D}^+ + \text{HD} \\ \overset{(b)}{\longrightarrow} \ \text{HD}^+ + \text{D} \\ \overset{(c)}{\longrightarrow} \ \text{D}_2^+ + \text{H} \end{array} \qquad (8.31)
$$

Accurate proton beam scattering experiments have yielded total cross sections for each of the three reactions shown in Eqn. 8.31, as plotted in Fig. 8.21. Using an *ab initio* set of 'interesting' potential energy surfaces, quasiclassical trajectories, and a so-called surface-hopping variant of the LZ curve-crossing approximation, J. C. Tully and R. K. Preston carried out calculations of these cross sections. Results are plotted as the points in Fig. 8.21. It is seen that they are in quite good agreement with experiment, giving some confidence to the

---

$\longrightarrow$

FIG. 8.22. Potential energy surfaces to model the reaction K + IR $\rightarrow$ KI + R (R $\equiv$ CH$_3$) plotted for four different configuration angles: (a) 45°, (b) 70°, (c) 90°, (d) 180°, as indicated. Contours in kcal mol^{-1}. These are pseudoadiabatic surfaces calculated by a modified DIM (diatomics-in-molecules) method; dashed curves indicate the locus of intersection of covalent with ionic surfaces. Adapted from R. A. LaBudde, P. J. Kuntz, R. B. Bernstein, and R. D. Levine, *J. Chem. Phys.* **59**, 6286 (1973); details therein.

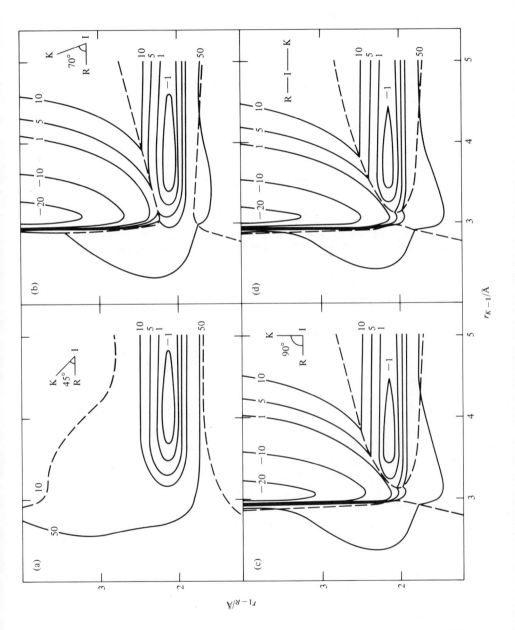

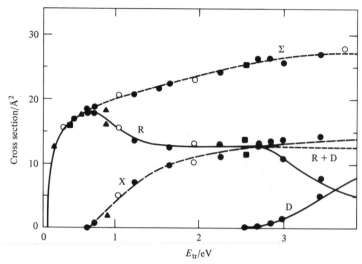

FIG. 8.23. Cross sections calculated via the quasiclassical trajectory method for the reactions shown in Eqn. 8.32, using the potential surface of Fig. 8.22. Symbols as follows: R, reaction to KI product; X, re-crossing back to entrance valley; D, collision-induced dissociation; Σ, sum of R, X, and D. Adapted from reference of Fig. 8.22.

mixed classical-quantal procedure developed to deal with nonadiabatic reactions for a three-atom system.

The surface-hopping approximation has been widely adopted to deal with polyatomic systems; unfortunately, the interaction matrix elements can only be guessed for most realistic systems and so there is limited predictive power.

One attempt to circumvent this problem was that by R. A. LaBudde, P. J. Kuntz, R. B. Bernstein, and R. D. Levine to deal with the $CH_3I + K$ reaction. First the six-atom system was replaced by a three-atom mimic, i.e., RI + K (where R is a 'point' methyl). Then the so-called diatomics-in-molecules (DIM) approximation was used to construct a set of diabatic potential surfaces for the system. Where diabatic surfaces intersected, the lower one was chosen. By this means, the lower 'pseudoadiabatic' surface was approximated. The 'seam' between the covalent surface (at large K–IR separations) and the ionic surface (for KI + R) was allowed to remain, producing a discontinuity in the forces at the seam (but otherwise no problem).

Figure 8.22 shows the resulting pseudoadiabatic surface, plotted for several angles of approach (as indicated). It is of parenthetical interest to note the strong orientation dependence which induces the reactive asymmetry (steric effect) mentioned in Chapter 6.

Quasiclassical trajectories were run on this pseudoadiabatic surface and cross sections computed for the processes:

$$K + IR \begin{cases} \xrightarrow{R} KI + R & \text{Reaction} \\ \xrightarrow{X} KIR \rightarrow K + IR & \text{Re-crossing back to entrance} \quad (8.32) \\ & \text{valley} \\ \xrightarrow{D} KIR \rightarrow K + I + R & \text{Collision-induced dissociation} \end{cases}$$

Plotted in Fig. 8.23 are the various cross sections for reaction (R), recrossing (X), dissociation (D), and their sum (Σ). To be noted is the appearance of a maximum in the reaction cross section followed by a decline (which is in qualitative accord with experiment).

Considerable theoretical-computational effort is currently being devoted to the problem of dealing with nonadiabatic processes involving more than two atoms. Great progress has been made by the FOM group in an attempt to explain a large body of experimental data on the scattering of fast atomic and molecular beams. There is still much to be done, however, in this very interesting field.

Review literature

MOT33, MUS66, STE66, STE68, BER70, BER70a, MAS71, MAP72, CHI74, LEV74, NIK74, NIK74a, BAE75, KEM75, MCG75, STE75, CHI76, TUL76, TUL77, TUL77a, WIL77, ZEW78, BER79, CHI79, DAV79, DIE79, FAR79, HER79, LOS79, WEI79, SMI80, JOR81, MCG81

9 Information-theoretic approach to the analysis of state-to-state chemical dynamics

9.1. Information theory applied to state-to-state dynamics

The formal link between statistical mechanics and information theory (SHA49) was made by E. T. Jaynes some 25 years ago. The information-theoretic (IT) approach was first applied to the field of molecular reaction dynamics by R. D. Levine, R. B. Bernstein, and co-workers A. Ben-Shaul, B. R. Johnson, and others a decade ago. It was originally directed towards the codification, compaction, and correlation of experimental (and computer-simulated) data on reactive molecular scattering, and later extended to deal with inelastic and elastic scattering and, most important, with state-to-state chemistry. A number of excellent reviews of the IT approach to chemical dynamics have appeared (see review literature), and so the present chapter will present only a small sampling of the applications relevant to state-to-state dynamics.

It is clear from Chapters 2, 6, and 7 that elementary chemical reactions are characterised by *specificity* in the energy disposal and *selectivity* of energy consumption. The question of the role of the individual quantum state of the reagent in the formation of a given quantum state of the product is a central one in state-to-state molecular reaction dynamics.

One of the goals of the IT approach is to systematize the large volume of data on state-to-state cross sections (and rate coefficients) for a given chemical reaction. The magnitude of the problem is best understood by consideration of the matrix of cross sections $\sigma_{fi}(E)$ from a given set of reagent states i to product states f. This cross section matrix at a given total energy E is spelled out in Table 9.1.

We define the probability of a given final state to be

$$P(f) = \sigma_{fi}/\sigma_i \qquad (9.1)$$

where $\sigma_i = \Sigma_f \sigma_{fi}$ at the given total energy E.

In a general sense, let $P(a)$ be the probability of a given outcome, say, a, of the reactive collision. We define a prior-expectation probability $P^0(a)$ taking account of the knowledge that certain outcomes are intrinsically more probable than others. (As an example, energy levels which are g-fold degenerate have an inherent g-fold factor higher probability of population.) Our definition of the 'prior' is the particular distribution of outcomes that is the most entropic (thus the least informative). It is sometimes called the 'statistical' distribution.

TABLE 1
Cross section matrix (BER75)

		Reactant state i				
		1	2	3	...	$N(E)$
Product state f	1	σ_{11}	σ_{12}	σ_{13}	...	σ_{1N}
	2	σ_{21}	σ_{22}	σ_{23}	...	σ_{2N}
	3	σ_{31}	σ_{32}	σ_{33}	...	σ_{3N}
	.	.	.	.		.
	.	.	.	.		.
	.	.	.	.		.
	$N'(E)$	$\sigma_{N'1}$	$\sigma_{N'2}$	$\sigma_{N'3}$	...	$\sigma_{N'N}$

All probability density functions are normalized, so that

$$\sum_a P(a) = \sum_a P^0(a) = 1 \tag{9.2}$$

9.2. The surprisal and the entropy deficiency

A measure of the amount of information provided by a knowledge of $P(a)$ is $I(a)$, the quantity termed the 'surprisal' of the outcome a, defined

$$I(a) = -\ln[P(a)/P^0(a)] \tag{9.3}$$

The surprisal is a direct measure of the deviance from prior expectation of outcome a.

The average of the surprisal (over all possible outcomes) is the amount of information obtained from a knowledge of the full set of probabilities $P(a)$; it is the 'information content' I of the entire distribution, given by

$$I = \sum_a P(a)\ln[P(a)/P^0(a)] \tag{9.4}$$

In the limit when the observed distribution matches the prior expectation, the information content of the distribution is zero. Otherwise $I > 0$, i.e., the information content is always positive, as can be seen directly. Let $x = P^0/P$, so that $I = \Sigma(P^0/x)(-\ln x)$. Thus $-I = \Sigma P^0(\ln x)/x$. But $\ln x < x - 1$ (except for $x = 1$), and so $-I < \Sigma P^0[1 - (1/x)] = \Sigma(P^0 - P) = \Sigma P^0 - \Sigma P = 1 - 1 = 0$. In other words, $I > 0$ except when $P(a) = P^0(a)$.

The entropy of a distribution is closely related to the information content. (Recall the conventional definition of the entropy S in molar units, i.e., $S =$

$-R\Sigma_n \, P(n) \ln P(n)$, where $R = kN_{avog}$ is the gas constant.) From the above result, it is clear that the entropy of the observed distribution will be smaller than that of the prior (the most entropic). The amount of the difference is known as the 'entropy deficiency' DS, i.e.,

$$DS = RI \geq 0 \tag{9.5}$$

with the entropy deficiency expressed in conventional molar entropy units, e.u.). The deficiency DS is the difference between the maximum and the actual value of the entropy of the distribution.

A large entropy deficiency (always nonnegative, of course) implies a high degree of specificity of the distribution $P(a)$, i.e., strong 'preference' for certain outcomes and thus very 'nonstatistical' product state populations. The entropy deficiency can be thought of as a *global* measurement of the state-to-state specificity of the reaction, whereas the surprisal $I(a)$ is a *local* measure of the specificity of a particular outcome a. Similar considerations apply to the problem of the selectivity of utilization of reagent states. Using the principle of microscopic reversibility, we can relate the surprisal of the product state distribution for an exoergic reaction with that of the reagent state consumption for the inverse, endoergic reaction.

9.3. Surprisal of product state distributions

Returning to the surprisal concept as it pertains to product state distributions, let us first have a look at experimental surprisal analyses of the vibrational populations of products from several elementary atom-molecule reactions, as first carried out by A. Ben-Shaul, R. D. Levine, and R. B. Bernstein. Figure 9.1 shows results for the exoergic reaction

$$F + HBr \rightarrow HF(v') + Br \tag{9.6}$$

plotted as a function of $f_{v'}$, the fraction of the total available energy E which appears as product vibration.

In the lower panel of Fig. 9.1, the relative populations of the HF vibrational states, $P(v')$, are compared with prior expectation, $P^0(v')$, the latter based on calculations to be described below. Clearly, the populations of the higher vibrational states are enhanced ('population inversion') relative to expectation on the basis of a statistical distribution. The upper panel shows a surprisal plot; the vibrational surprisal turns out to be a near-linear function of $f_{v'}$ with a slope $\lambda = -4$. (The parameter λ is negative when the system exhibits vibrational population inversion.)

Figure 9.2 shows analogous results for the reaction

$$O + CS \rightarrow CO(v') + S \tag{9.7}$$

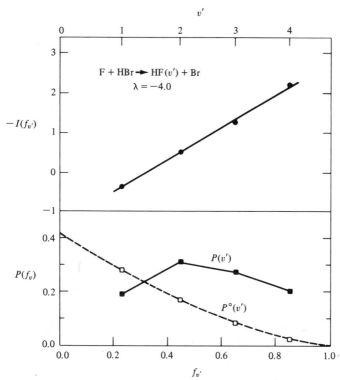

FIG. 9.1. Products' vibrational state distribution for the reaction of F + HBr. *Lower:* solid squares, probability P of formation of HF in various indicated states v' (corresponding to a fraction $f_{v'}$ of the available energy going into product vibration); open squares, prior-expectation probability P^0. *Upper:* surprisal plot of same data, $-I(f_{v'}) \equiv \ln[P(f_{v'})/P_{v'}^0]$. Slope yields vibrational surprisal parameter $\lambda = -4.0$. Adapted from A. Ben-Shaul, *Chem. Phys.* **1**, 244 (1973); data source therein. See also LEV74a.

Here the population inversion is even more pronounced, and correspondingly the value of the vibrational surprisal parameter is more negative (here $\lambda = -7.7$).

Figure 9.3 summarizes vibrational state distributions for the products of two isotopic reactions

$$\text{Cl} + \begin{cases} \text{HI} \\ \\ \text{DI} \end{cases} \rightarrow \begin{cases} \text{HCl}(v') \\ \\ \text{DCl}(v') \end{cases} + \text{I} \tag{9.8}$$

Here it is found that the reduced variable $f_{v'}$ serves to unify the data for the two isotopic reactions, i.e., there is essentially a single functionality $P(f_{v'})$ and a single vibrational surprisal for the isotopic reactions. Once again, as in Figs.

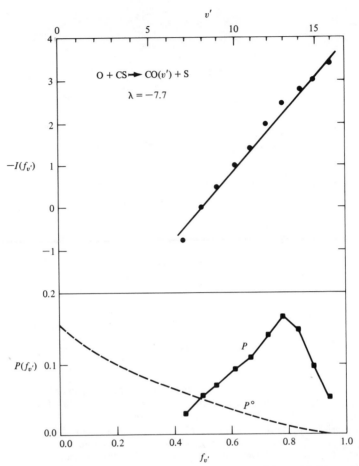

FIG. 9.2. Plots similar to those of Fig. 9.1 for the reaction of O + CS. From the ratio of P to P^0, the vibrational surprisal $I(f_{v'})$ is found. The upper graph is a surprisal plot yielding $\lambda = -7.7$. Adapted from reference of Fig. 9.1; data source therein. See also LEV74a.

9.1 and 9.2, the surprisal plot yields a near-linear representation of the experimental data. Figure 9.4 shows data and surprisal plots (for the same reaction) over a range of total energies. Figure 9.5 shows similar results for the isotopic reactions

$$F + \begin{cases} H_2 \\ D_2 \\ HD \\ DH \end{cases} \rightarrow \begin{cases} HF + H \\ DF + D \\ HF + D \\ DF + H \end{cases} \qquad (9.9)$$

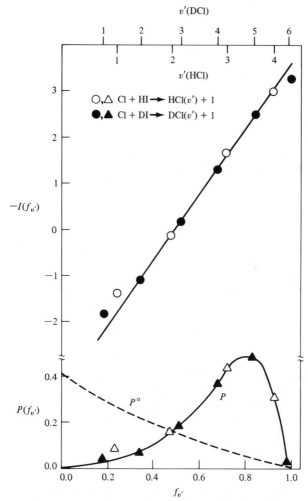

FIG. 9.3. Plots similar to those of Figs. 9.1 and 9.2 for the isotopic reactions Cl + HI, DI. Open symbols denote HCl product, solid symbols DCl. Note that the use of the reduced energy variable $f_{v'}$ unifies the data for the two isotopic product molecules. The upper graph is a vibrational surprisal plot with $\lambda = -8.0$. Adapted from A. Ben-Shaul, R. D. Levine, and R. B. Bernstein, *Chem. Phys. Lett.* **15**, 160 (1972); *J. Chem. Phys.* **57**, 5427 (1972); data sources therein.

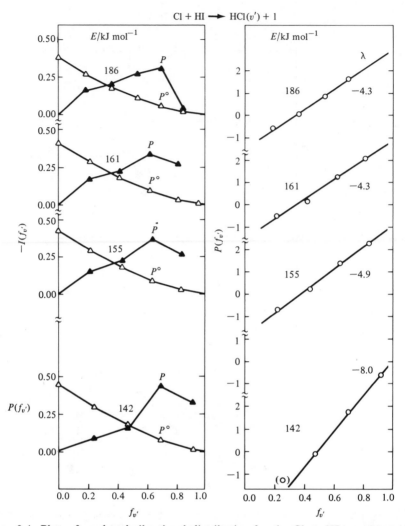

FIG. 9.4. Plots of products' vibrational distribution for the Cl + HI reaction as a function of the total energy E of the system. *Left:* P and P^0 versus $f_{v'}$ at the indicated collision energies; *right:* vibrational surprisal plots of the same data, showing systematic decrease in the magnitude of λ as E increases. Adapted from A. Ben-Shaul, R. D. Levine, and R. B. Bernstein, *J. Chem. Phys.* **57**, 5427 (1972); details therein. Based on data of L. T. Cowley, D. S. Horne, and J. C. Polanyi, *Chem. Phys. Lett.* **12**, 144 (1971) and D. H. Maylotte, J. C. Polanyi, and K. B. Woodall, *J. Chem. Phys.* **57**, 1547 (1972).

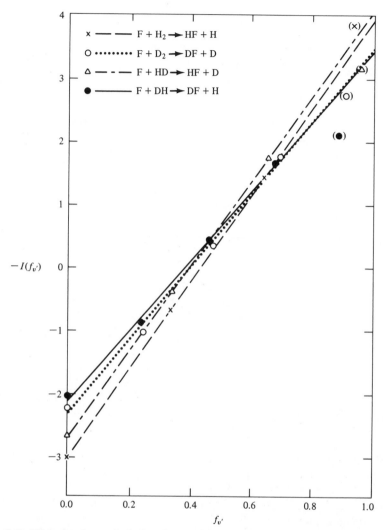

FIG. 9.5. Vibrational surprisal plots for the reactions of F atoms with isotopic hydrogen molecules. Small but significant differences in λ are found (the range is from −5.5 to −6.9). Adapted from M. J. Berry, *J. Chem. Phys.* **59**, 6229 (1973).

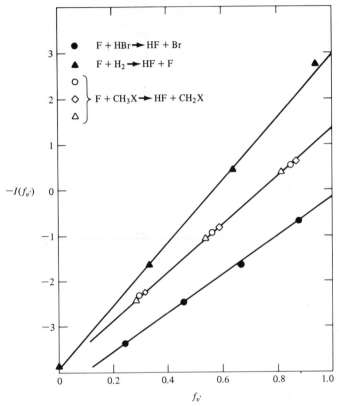

FIG. 9.6. Vibrational surprisal plots for several reactions of F atoms. Adapted from Fig. 2.10 of BEN81; data sources therein.

Although the individual surprisal plots are fairly linear, small but significant differences among the isotopic reactions are noted.

Figure 9.6 is a comparison of vibrational surprisal plots for several exoergic reactions forming HF product. The F + H$_2$ reaction produces the most extensive population inversion.

For many elementary reaction systems, it has been found that surprisals (both vibrational and rotational) are smooth, simple functions of the appropriate reduced energy variables. For example, the vibrational surprisals for dozens of elementary exoergic reactions are well represented by the simple linear equation

$$I(f_{v'}) = \lambda_0 + \lambda f_{v'} \tag{9.10}$$

where the intercept constant λ_0 is governed by considerations of normalization of the distribution. The origin of such simple behaviour (which stems from the so-called maximum entropy principle) will be discussed below, after we have dealt with the more practical problem of ascertaining the prior distribution (P^0) against which to compare the experimental one (P).

9.4. The prior-expectation distribution

For reactions that populate the various internal states of the products according to statistical or phase space considerations alone (i.e., no dynamic effects), the product state distribution and its dependence on reagent states and upon total energy E are the properties of the 'prior.' Thus before evaluating the surprisal and the entropy deficiency, it is necessary to calculate this 'statistical' (or 'phase space,' or 'density of states') prior distribution. This requires only quantum state-counting, as pointed out by J. L. Kinsey, since the prior expectation is to populate every energetically allowed group of product states with a probability proportional to the number of such states in that group.

First we note that the density of translational states is given by the expression

$$\rho_{tr}(E_{tr}) = \frac{\mu^{3/2}}{2^{1/2}\pi^2 \hbar^3} E_{tr}^{1/2} \tag{9.11}$$

with ρ_{tr} the number of states per unit volume per unit interval in E_{tr}. Following procedures outlined in the references, expressions have been obtained for the *total* density of states $\rho(E)$ (including internal as well as translational energy states) for various cases, i.e., combinations of product species. In the simplest case, where the product set is a diatomic plus atom, the total density of product states is

$$\rho(E) = \sum_{v'=0}^{v'^*} \sum_{J'=0}^{J'^*(v')} (2J' + 1)\rho_{tr}(E - E'_{int}) \tag{9.12}$$

where $J'^*(v')$ is the largest energetically allowed J' for the given v', and v'^* the largest v' allowed for given E. Here $E - E'_{int} = E'_{tr}$, i.e., $\rho_{tr}(E'_{tr}) \equiv \rho_{tr}(E - E'_{int})$.

Thus the prior-expectation probability of individual vibrational states, $P^0(v', J')$, is

$$P^0(v', J') = (2J + 1)\rho_{tr}(E - E'_{int})/\rho(E) \tag{9.13}$$

where the denominator is the $\rho(E)$ of Eqn. 9.11, serving as the normalizing factor. The prior distribution of vibrational states (*irrespective* of J'), $P^0(v')$, is therefore the sum over J'

$$P^0(v') = \sum_{J=0}^{J^*(v')} P^0(v', J') \tag{9.14a}$$

For a given vibrational state v', the rotational prior is

$$P^0(J') \propto (2J' + 1)(E - E_{v'} - E_J)^{1/2} \qquad (9.14b)$$

In the 'ideal' limiting case in which the diatomic energy levels are approximated by the rigid-rotor harmonic oscillator (RRHO) level scheme, simple formulas for the priors have been worked out in terms of the reduced energy variables. One obtains compact approximate (RRHO) expressions, such as

$$P^0(f_v) = \tfrac{5}{2}(1 - f_v)^{3/2} \qquad (9.15a)$$

$$P^0(f_T) = \tfrac{15}{4}f_T^{1/2}(1 - f_T) \qquad (9.15b)$$

$$P^0(f_R|f_v) = \frac{3}{2}\frac{(1 - f_v - f_R)^{1/2}}{(1 - f_v)^{3/2}} = \frac{3}{2}(1 - g)^{1/2}/(1 - f_v) \qquad (9.15c)$$

where we have dropped the primes. The reduced rotational energy variable g is defined implicitly in Eqn. 9.15c, i.e., $g = f_R/(1 - f_v)$.

Thus the vibrational surprisal (for the RRHO case) is given by

$$I(f_v) = -\ln[P(f_v)/P^0(f_v)] = -\ln\left[\frac{2P(f_v)}{5(1 - f_v)^{3/2}}\right] \qquad (9.16)$$

For more general situations, such as those for atom-polyatomic or diatomic-polyatomic product pairs, various approximate expressions for the different priors (at constant E) have been reported. They are summarized in LEV79. Analogous prior state-to-state rate coefficients at constant *temperature* (rather than constant E) have been worked out.

9.5. 'Prior' rate coefficients and surprisal

Figure 7.6 displayed one such result, that for the atom-diatomic (RRHO) system, which is quite instructive when dealing with the question of 'vibrational enhancement' of endoergic reactions. (Refer to Chapter 7 for further discussion of the prior rate constant at temperature T as a function of the dimensionless variable $(E_v - \Delta E_0)/kT$. It was found by R. D. Levine and J. Manz that in the 'strongly endothermic' limit, where $E_v - \Delta E_0 \ll 0$, the prior rate can be shown to be of the form

$$k_v^0(T) \approx C(T)\exp[E_v - \Delta E_0)/kT] \qquad (9.17)$$

where $C(T)$ is only weakly dependent upon temperature. In this limit,

$$(\partial \ln k_v^0/\partial E_v)_T = \frac{1}{kT} \qquad (9.18)$$

With added reagent vibrational energy, the differential vibrational enhancement is reduced; in the 'exothermic limit,'

$$k_v^0(T) \propto [(E_{int} - \Delta E_0)/kT]^{5/2} \tag{9.19}$$

The prior rate is independent of the magnitude of any intrinsic activation barrier for the reaction and depends only upon the reduced variable $(E_v - \Delta E_0)/kT$.

The vibrational surprisal of the rate constant expresses the deviance of the rate constant from state v (at the given temperature) from its prior expectation. It is defined

$$I_v(T) \equiv -\ln[k_v(T)/k_v^0(T)] \tag{9.20}$$

In many cases, the surprisal is found to be essentially linear and can be expressed in the simple form

$$I_v = I_0 + \lambda E_v/kT \tag{9.21}$$

In the strongly endothermic limit, this yields (from Eqn. 9.16) the following approximate expression for the rate constant:

$$k_v(T) = A \exp\{-[\Delta E_0 - (1 - \lambda)E_v]/kT\} \tag{9.22}$$

so that

$$(\partial \ln k/\partial E_v)_T = (1 - \lambda)/kT \tag{9.23}$$

Comparing this result with Eqn. 7.6, one can identify the empirical parameter α with $1 - \lambda$. For many systems the λ is negative, lying in the range $0 \geq \lambda \geq -0.25$. This implies an *extra* efficacy of vibrational excitation over that of the prior. This is related to the frequent occurrence of high vibrational excitation of nascent products in the exoergic reactions (recall the discussion of Chapter 2 on microreversibility). The IT approach serves to identify that part of the vibrational enhancement due to the dynamic factors, as distinguished from that following from the statistical prior. As usual, the surprisal refers to the deviance per se from prior expectation.

The problem of the dependence upon total energy E of the state-to-state rate constant (and its surprisal) is a subject of considerable interest.

An interesting example of a 'nonsurprising' energy dependence comes from experiments on various chemiluminescent oxidation reactions of La and Sc, e.g.,

$$
\begin{array}{c}
\text{La} + \text{O}_2 \rightarrow \text{LaO*(A,B,C)} + \text{O} \\
\Big\downarrow{\scriptstyle h\nu} \\
\text{LaO(X)}
\end{array}
\tag{9.24}
$$

From the chemiluminescence, it was possible to deduce relative rates of formation of each of three excited electronic states of the LaO* product. The results are shown in Fig. 9.7. The experimental relative rates agree well with the priors (the solid curves) calculated from the product density of states.

9.6. Maximum entropy principle applied to rotational state distributions

Next let us try to rationalize the simple, near-linear surprisal behaviour observed for so many elementary reactions. The route is the so-called maximum entropy principle, which states that the most likely distribution is the one with the maximum entropy, subject to the known constraints on the system. We shall now exemplify the procedure with some simple illustrations. Consider the rotational excitation of a rigid, $j = 0$, rotor by a spherical atom at given E. The postcollision rotational state distribution is designated (dropping primes) $P(j)$ for $j = 0,1,2,\cdots, j_{max}(E)$. The information content of the $P(j)$ distribution is (in accord with Eqn. 9.4)

$$I(E) = \sum_{j=0}^{j_{max}} P(j)\ln[P(j)/P^0(j)] \qquad (9.25)$$

where $P^0(j)$ is the prior expectation given by

$$P^0(j) = (2j + 1)(E - E_j)^{1/2}/ \sum_{j=0}^{j_{max}} (2j + 1)(E - E_j)^{1/2} \qquad (9.26)$$

The normalizations are

$$\sum_{j=0}^{j_{max}} P(j) = \sum_{j=0}^{j_{max}} P^0(j) = 1 \qquad (9.27)$$

We must maximize the entropy of the distribution $P(j)$ subject to the known constraints, i.e., (a) normalization and (b) an as yet unknown average value of the final rotational energy, say $<E_j>$. Maximizing S is equivalent to minimizing I (the information content), subject to the same constraints. The method is straightforward (LEV76). We must find the simple minimum of a Lagrangian L, written

$$L = I + \alpha<1> + \beta<E_j> \qquad (9.28)$$

Here α and β are the Lagrange parameters, to be determined from the variational condition

$$\delta L = 0 \qquad (9.29)$$

Thus

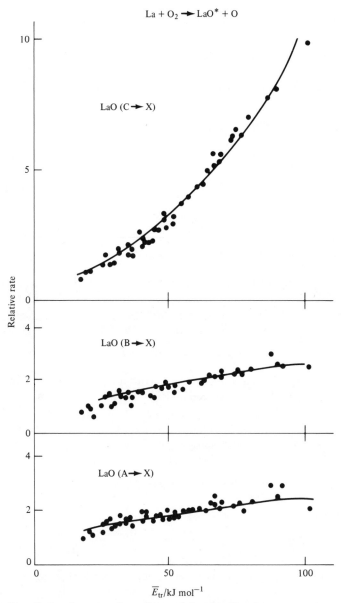

FIG. 9.7. Translational energy dependence of the chemiluminescence cross section for the three concurrent reactions of La with O_2 (forming LaO in indicated electronic states). *Points:* experimental; *solid curves:* theoretical, i.e., information-theoretic prior-expectation calculation. Adapted from D. M. Manos and J. M. Parson, *J. Chem. Phys.* **69**, 231 (1978); details therein.

$$0 = \delta L = \delta \left(\sum_j P_j \ln\frac{P_j}{P_j^0} + \alpha <1> + \beta <E_j> \right)$$

$$= \delta \left[\sum_j \left(P_j \ln P_j - P_j \ln P_j^0 + \alpha P_j + \beta P_j E_j \right) \right]$$

$$= \sum_j \left(P_j \, \delta \ln P_j + \ln P_j \, \delta P_j - \ln P_j^0 \, \delta P_j + \alpha \delta P_j + \beta E_j \, \delta P_j \right) \tag{9.30}$$

$$= \sum_j \delta P_j \left(1 + \ln\frac{P_j}{P_j^0} + \alpha + \beta E_j \right)$$

Since δP_j's are arbitrary,

$$1 + \ln\frac{P_j}{P_j^0} + \alpha + \beta E_j = 0 \tag{9.31}$$

Thus

$$-\ln(P_j/P_j^0) = I(j) = (1 + \alpha) + \beta E_j = \gamma + \beta E_j \tag{9.32}$$

Equation 9.32 implies that a plot of I versus E_j will be linear, i.e., a 'linear surprisal.' Its slope β turns out to be approximately $[<E_j>]^{-1}$, BER76. (Note that $\beta \neq (kT)^{-1}$; there is no 'temperature' here!)

The entropy deficiency of $P(j)$ is (using Eqn. 9.4)

$$DS = -R(\gamma + \beta <E_j>) \tag{9.33}$$

The more general form of Eqn. 9.32 is written in terms of the reduced rotational energy variable g, introduced by R. D. Levine, B. R. Johnson, and R. B. Bernstein. It is defined by Eqn. 9.15c, i.e., $g \equiv E_{rot}(J')/E_{rot}(J'_{max})$, where $E_{rot}(J'_{max})$ is the maximum rotational energy available for the product, dependent upon the product's vibrational state: $E_{rot}(J'_{max}) = E - E_{v'}$. (Obviously, for $v' = 0$, $E_{rot}(\quad) = E$.) This result can be written

$$I(j \rightarrow j') = \theta_0 + \theta_R |\Delta g| \tag{9.34}$$

where

$$\Delta g = g' - g = \frac{\Delta E_{rot}}{E - E_{v'}} = \frac{E_{rot}(J') - E_{rot}(J)}{E - E_{v'}} \tag{9.35}$$

and θ_R is the rotational surprisal parameter. (Note that θ_R is *not* equal to $(kT_{rot})^{-1}$ since there is *no* well-defined T_{rot} concept for a disequilibrium rotational distribution.) We shall return shortly to illustrate the practical application of Eqns. 9.32 and 9.34, but first we shall work through another example of the maximum entropy procedure. Here we shall apply the method to vibrational populations but take into consideration *more* than the first moment constraint.

Assume an observed set of populations $P(v')$. As before, we find it advantageous to work in 'reduced variable space': $f_{v'}$ is the fraction of the total available energy in product vibration, i.e.,

$$f_{v'} = \frac{E_{vib}(v')}{E} \approx \frac{v'\hbar\omega}{E} \qquad 0 \leq f_{v'} \leq 1 \qquad (9.36)$$

Dropping primes, the vibrational surprisal is, as usual,

$$I(f_v) = -\ln[P(f_v)/P^0(f_v)] \qquad (9.37)$$

Suppose inspection of the experimental results, i.e., $P(f_v)$, shows that they contain more than one independent moment, i.e., that the distribution can be characterised by, say, two independent moments, $<f_v>$ and $<f_v^2>$, with all higher moments being either zero or undetermined within the experimental error.

Then the maximum entropy principle tells us that $P(f_v)$ is that distribution which minimizes I (or DS) subject to the three independent constraints:

$$1 = \sum_v P(v) \qquad \text{normalization} \qquad (9.38a)$$

$$<f_v> = \Sigma P(v)f_v \qquad \text{first moment given} \qquad (9.38b)$$

$$<f_v^2> = \Sigma P(v)f_v^2 \qquad \text{second moment given} \qquad (9.38c)$$

As before, we construct the Lagrangian L and minimize:

$$L = I + (\lambda_0 - 1)<1> + \lambda_1 <f_v> + \lambda_2 <f_v^2> \qquad (9.39)$$

Then

$$0 = \delta L = \delta \left[\sum_v P_v \ln\left(\frac{P_v}{P_v^0}\right) + (\lambda_0 - 1)<1> \right.$$

$$\left. + \lambda_1 <f_v> + \lambda_v <f_v^2> \right] \qquad (9.40)$$

$$= \sum_v \delta P_v \left[1 + \ln\left(\frac{P_v}{P_v^0}\right) + (\lambda - 1) + \lambda_1 f_v + \lambda_2 f_v^2 \right]$$

Therefore

$$-\ln\left(\frac{P_v}{P_v^0}\right) \equiv I(f_v) = \lambda_0 + \lambda_1 f_v + \lambda_2 f_v^2 \qquad (9.41)$$

i.e., a quadratic (nonlinear) vibrational surprisal. If all we know about the distribution is its first moment $<f_v>$, the above simplifies to the common linear surprisal limit

$$I(f_v) = \lambda_0 + \lambda f_v \qquad (9.42)$$

so that

$$P_v = P_v^0 \exp[-I(f_v)] = P_v^0 e^{-\lambda_v}/Q \tag{9.43}$$

where

$$Q = e^{\lambda_0} = \sum_v P_v^0 e^{-\lambda_v} \tag{9.44}$$

is a partition function (from the normalization). For more on the theory, the review articles should be consulted.

Now we shall show some examples of rotational (and vibrational) surprisal analysis of both theoretical-computational and experimental data.

Figure 9.8 is a rotational surprisal plot for the product of the $H + H_2$ exchange reaction, i.e.,

$$H + H_2(v = 0, J = 0) \rightarrow H_2(v' = 0, J') + H \tag{9.45}$$

The $P(J')$ points are *ab initio*, from a full 3-D quantal computation by R. E. Wyatt and A. B. Elkowitz on an *ab initio*–calculated H_3 potential energy surface. A linear rotational surprisal is found.

Figure 9.9 summarizes similar (but quite independent) *ab initio* calculations at several total energies. The surprisal parameter shows a slight energy dependence.

For experiments on rotational energy transfer, surprisal analysis has been widely used to compact large quantities of state-to-state rate and/or cross section data. The approach has been especially appropriate in dealing with the output of theoretical computations of rotational transition cross sections, where there is usually an 'embarrassment of riches' in terms of numerical results (envisage Table 9.1 with $N \approx N' \approx 10^2$, i.e., 10^4 entries!). This is illustrated by various calculations (somewhat smaller in scope, however) on the Ar-N_2 system, using different assumed (empirical) anisotropic potential energy surfaces, typified by those shown in Fig. 9.10.

In one computational study using the full close-coupled quantal scattering treatment, cross sections for a wide variety of rotational transitions starting from many different initial J states were computed at several values of total energy E. Results at a given E could be brought together, i.e., compacted, by a surprisal plot, as shown in Fig. 9.11.

Using the same potential surface, quasiclassical trajectory calculations were carried out in an attempt (fairly successful) to simulate the quantal-computed cross sections. These results are shown in the form of a surprisal plot in Fig. 9.12. Figure 9.13 is a similar representation of computational results for the He-H_2 system.

With the advent of experimental data by laser pumping and fluorescence techniques, it has been possible to observe very large collision-induced changes in rotational quantum numbers, even though such transitions are characterised

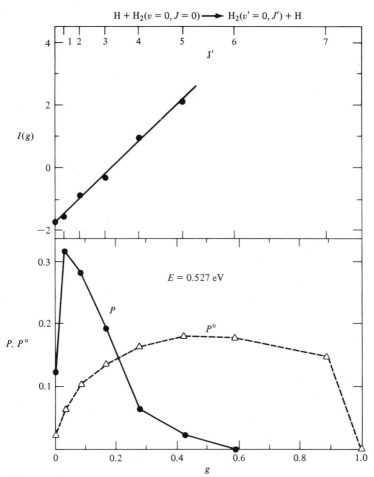

FIG. 9.8. Plots of rotational state distribution for the $H_2(v' = 0)$ product of the hydrogen exchange reaction $H + H_2(v = 0, J = 0) \rightarrow H_2(v' = 0, J) + H$ at the indicated total energy. *Lower:* solid circles, $P(J')$ calculated via accurate quantal method; open triangles, prior $P^0(J')$. *Upper:* rotational surprisal plot, all as a function of the products' reduced rotational energy g. Adapted from R. E. Wyatt, *Chem. Phys. Lett.* **34**, 167 (1975); details therein.

by very small cross sections. It is found that the surprisal plots I versus $|\Delta g|$ are somewhat nonlinear and that a power law, as proposed by D. E. Pritchard et al., is a better representation of the data (i.e., a better independent variable is $|\Delta g|^\gamma$, where γ is a constant near but not necessarily equal to unity).

As more experimental state-to-state data accumulate and the precision of results becomes greater, it is becoming clear that nonlinear surprisals are more common and that multiple constraints are needed to account for the results.

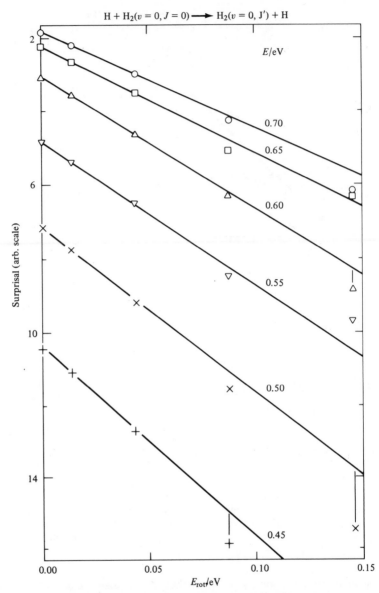

FIG. 9.9. Rotational surprisal plots of the reaction of Fig. 9.8 based on quantal computations at several indicated values of the total energy. (Results are consistent with those of Fig. 9.8; difference in convention of ordinate and abscissa.) Adapted from A. Kuppermann and G. C. Schatz, *J. Chem. Phys.* **65**, 4668 (1976); see also R. B. Bernstein, BER76; details therein.

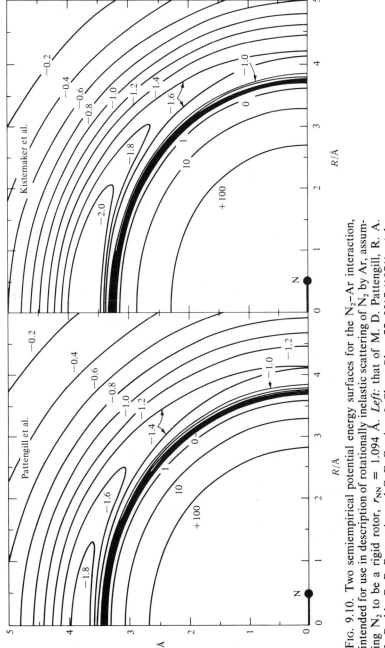

FIG. 9.10. Two semiempirical potential energy surfaces for the N_2–Ar interaction, intended for use in description of rotationally inelastic scattering of N_2 by Ar, assuming N_2 to be a rigid rotor, r_{NN} = 1.094 Å. *Left*: that of M. D. Pattengill, R. A. LaBudde, R. B. Bernstein, and C. F. Curtiss, *J. Chem. Phys.* **55**, 5517 (1971); *right*: that of P. Kistemaker and A. E. DeVries, *Chem. Phys.* **7**, 371 (1975). Energy contours in units of 10^{-21} J. Inelastic scattering cross sections calculated (via the quasiclassical trajectory method) for these surfaces, as well as rotational surprisals, have been compared by M. D. Pattengill and R. B. Bernstein, *J. Chem. Phys.* **65**, 4007 (1976); details therein.

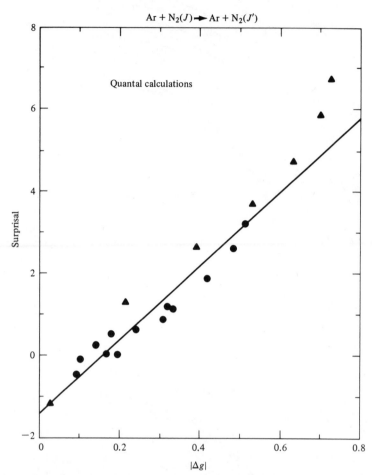

FIG. 9.11. Rotational surprisal plot, adapted from R. D. Levine, R. B. Bernstein, P. Kahana, I. Procaccia, and E. T. Upchurch, *J. Chem. Phys.* **64**, 796 (1976), of all the energetically allowed state-to-state rotational transition cross sections for the Ar–N_2 system computed quantum-mechanically by R. T. Pack, *J. Chem. Phys.* **62**, 3143 (1975), based upon the potential surface of Pattengill et al. (reference of Fig. 9.10). Energy 0.061 eV. Circles denote fully converged accurate cross sections; triangles, less accurate.

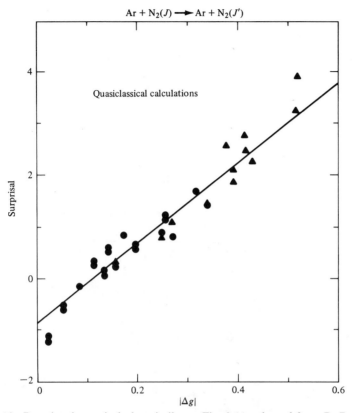

$$Ar + N_2(J) \longrightarrow Ar + N_2(J')$$

Quasiclassical calculations

FIG. 9.12. Rotational surprisal plot, similar to Fig. 9.11, adapted from R. D. Levine et al., reference of Fig. 9.11, of cross sections for the same system computed quasi-classically by M. D. Pattengill, *Chem. Phys. Lett.* **36**, 25 (1975). Energy same as for Fig. 9.11. Symbols have same meaning. See R. D. Levine et al., reference of Fig. 9.11, for details.

This takes away from the simplicity of the IT approach but is the price one must pay, in effect, for more informative (and extensive) dynamic data.

For recent extensive treatment of IT as applied to state-to-state data, see BEN81.

9.7. Detailed application: a practical illustration

Let us return to a final 'practical example,' namely, the IT analysis of the chemiluminescent ion-molecule reaction

$$C^+(^2P) + \begin{cases} H_2 \\ D_2 \end{cases} \rightarrow \begin{cases} CH^+(A^1\Pi) + H \\ CD^+(A^1\Pi) + D \end{cases} \quad (9.46)$$

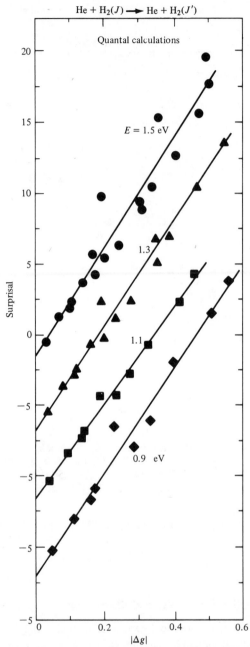

FIG. 9.13. Rotational surprisal plots for the He + H₂(J) model system at several indicated values of total energy, based upon quantal computations of H. A. Rabitz and G. Zarur, *J. Chem. Phys.* **61**, 5076 (1974), adapted from R. D. Levine et al., reference of Fig. 9.11; details therein.

Such experiments using energy-selected ion beams have been carried out in the laboratories of C. Ottinger, of J. J. Leventhal, and others. Measurements of the A → X fluorescence of the CH^+, CD^+ products have yielded rovibrational populations of the nascent A-state ions for experiments at several collision energies from threshold to 8 eV.

Figure 9.14 (upper panel) displays the original data points for both isotopic reactions at 4 eV in the form $P(J|v = 0)$ versus J. The solid lines are fits to the data assuming linear rotational surprisals (see below).

The top portion of the lower panel plots the same $P(J)$ versus g as abscissa (for CH^+ only), comparing the experimental results with the prior $P^0(J)$. Clearly, the observed rotational energy distribution is 'cold' with respect to the prior, peaking at $J = 11$ (versus 14 or 15 for the prior). The bottom portion of the lower panel is a plot of the closely related probability $P(g)$ versus g, where

$$P(g) = P(J) \frac{dJ}{dg} \propto \frac{P(J)}{(2J + 1)} \tag{9.47}$$

The results, in the form of $P(g, v = 0)$, are found experimentally to be the same for both isotopic products (cf. the near-isotopic invariance of most $P(f_v)$ data, such as for Reactions 9.8 and 9.9). [In converting $P(J)$ to $P(g)$, the Jacobian dJ/dg in Eqn. 9.47 has removed the 'trivial maximum' due to the $(2J + 1)$ degeneracy factor.] Also shown is the rigid-rotor prior

$$P^0(g) = 3/2(1 - g)^{1/2} \qquad (f_v = 0 \text{ case}) \tag{9.48}$$

Clearly, the observed $P(g)$ is a 'colder' distribution than the prior.

Figure 9.15 shows rotational surprisal plots $I(g) \equiv -\ln[P(g)/P^0(g)]$ for two experiments ($E = 5$ and 8 eV); the straight-line approximation is of the form

$$I = \theta_0 + \theta_R g \tag{9.49}$$

a special case of the general linear surprisal expression (Eqn. 9.34). Obviously, the data do not conform perfectly to Eqn. 9.49, but the deviations in the back-calculated $P(J)$ distributions (e.g., top panel of Fig. 9.14) are only slightly outside the experimental error limits. A fit with a two-constraint (quadratic) surprisal is, of course, better. The rotational surprisal parameters are isotope independent but vary with energy, decreasing from approximately 3 to nearly zero as E increases from 5 to 8 eV.

Next let us consider the vibrational distributions. Figure 9.16 displays the experimental data for CD^+ at $E = 6$, 7, and 8 eV. The $P(f_v)$ data points are compared with the prior

$$P^0(f_v) = (1 - f_v)^{3/2} / \sum_v (1 - f_v)^{3/2} \tag{9.50}$$

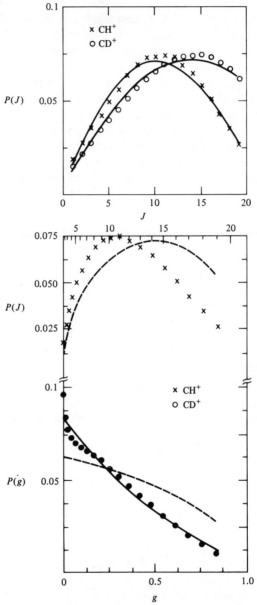

FIG. 9.14. Products' rotational state distributions from the isotopic chemiluminescent ion-molecule reactions of $C^+ + H_2$, D_2, from the experiments of I. Kusunoki and C. Ottinger, *J. Chem. Phys.* **71**, 4227 (1979); analyzed by E. Zamir, R. D. Levine, and R. B. Bernstein, *Chem. Phys.* **55**, 57 (1981). *Top panel:* $P(J)$ for CH^+ and CD^+ (sym-

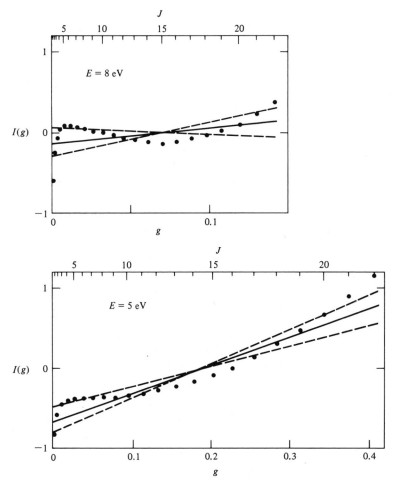

FIG. 9.15. Rotational surprisal plots of the 5- and 8-eV experiments of I. Kusunoki and C. Ottinger, reference of Fig. 9.14, adapted from E. Zamir et al., reference of Fig. 9.14; details therein.

boıs denote experiments at 4 eV); solid curves, calculated fits via information-theoretic method. *Bottom panel:* upper, $P(g)$ points for CH^+, compared with prior distribution (dashed curve); lower, $P(g)$ points for CH^+ and CD^+ (identical within experimental uncertainty), compared with calculated fit (solid curve) and the prior (dashed curve), $P^0(g) = \frac{3}{2}(1 - g)^{1/2}$. The calculated fits assumed a linear surprisal. See E. Zamir et al. for details.

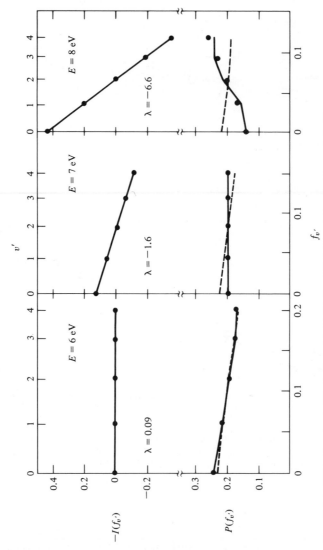

FIG. 9.16. Vibrational surprisal plots of $P(f_v)$ for CD^+ from the 6-, 7-, and 8-eV experiments of I. Kusunoki and C. Ottinger, adapted from E. Zamir et al., references of Fig. 9.14. *Lower:* $P(f_{v'})$ points experimental; dashed lines, prior $P^0(f_{v'})$. *Upper:* vibrational surprisal, lines with indicated values of λ.

and the vibrational surprisals $I(f_v) \equiv -\ln[P(f_v)/P^0(f_v)]$ plotted for each energy. They accord well with the linear surprisal form

$$P(f_v) = P^0(f_v)\exp[-(\lambda_0 + \lambda f_v)] \qquad (9.51)$$

and the fits to the original data with the best λ values from the surprisal slopes are shown in the lower portion of Fig. 9.16.

The energy dependence of the vibrational distributions is quite interesting. For $E \lesssim 6$ eV, $P(v) \approx P^0(v)$, i.e., $I(v) \approx 0$, the distributions are nonsurprising and accord well with the maximal entropy prior. However, as shown in Fig. 9.17, at higher E, population inversion sets in and λ values become negative. This is evidence for a gradual transition from a 'complex-mode' microscopic mechanism at low E to 'direct-mode' collision dynamics at higher collisions energies, as first noted in ion-molecule reactions by R. Wolfgang and A. Henglein and their co-workers. This behaviour is confirmed from ancillary data on translational and angular product distributions for such ion-molecule reactions, which proceed through an 'adduct complex' with an energy-dependent lifetime of the order of picoseconds. At higher translational energies, osculation becomes more and more evident (i.e., more forward scattering) and the direct (stripping) mode takes over, with concomitant vibrational excitation, population inversion, and strong negative surprisal parameters λ.

9.8. Concept of rotational temperature

One final topic is appropriate for discussion before leaving the subject of this chapter. It has to do with a common fallacy in the experimental literature on nascent rotational state distributions, especially for diatomic products from molecular or ion beam reactions. Perhaps this is a remnant from the 'old days' of the spectroscopy of flames, when vibrational and rotational temperatures were assigned on the basis of Boltzmann plots [e.g., for vibrations, $\ln P(v)$ versus $E_{\text{vib}} \approx hc\omega_e(v + \frac{1}{2})$, negative slope is $(kT_v)^{-1}$; for rotations, $\ln P(J)/(2J + 1)$ versus $E_{\text{rot}} \approx B_v hc(J^2 + J)$, negative of slope being $(kT_r)^{-1}$, as usual]. Here the concept of temperature may be appropriate since the nascent products can equilibrate via collisions with the 'heat bath.' And, of course, one always obtains positive T values!

However, in beam experiments at given total E (or with a narrow distribution of E), the nascent products' state distribution cannot be characterised by Boltzmann distribution since the available energy for the isolated system is fixed. Now in the case of vibrational state distributions, after the discovery of population inversion it was clear that the concept of a vibrational temperature (which would be negative from the positive slope of the Boltzmann plots) was inappropriate. For rotations, however, the plots may be linear over certain

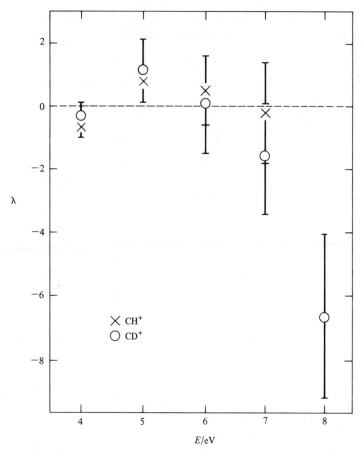

Fɪɢ. 9.17. Plot of energy dependence of vibrational surprisal parameters for CH^+ and CD^+, from the analysis by E. Zamir et al. of the data of I. Kusunoki and C. Ottinger, references of Fig. 9.14. Within the experimental uncertainty, there is no isotope effect. However, there is a significant trend towards negative λ for $E \geq 7$ eV. See E. Zamir et al. for discussion.

ranges of J and the slopes are indeed often found to be negative, leading to positive apparent rotational temperatures, sometimes quoted as such. Occasionally, these apparent T_{rot} values are very large (obviously, for a horizontal Boltzmann plot they become infinite!). Moreover, they differ for each vibrational manifold.

The information-theoretical approach starts by correctly limiting the energy range for both the prior $P^0(g)$ and the actual distribution $P(g)$ through the use of the reduced rotational energy g $[\equiv E_{rot}/(E - E_v)]$, which spans the proper range of E_{rot}. Then it compares $P(g)$ with $P^0(g)$ via the surprisal $I(g)$;

in the simplest case, the linear surprisal parameter θ_R is independent of v, and so one has characterised the entire rotational state distribution $P(J \mid v)$ by a single parameter. For this situation, apparent Boltzmann temperatures would show a predictable v dependence originating from the trivial contraction of the available energy span $(E - E_v)$ with increasing v. Were it not for the fact that for most systems studied, θ_R is positive (i.e., rotations colder than prior expectations, no population inversion), the use of T_{rot} for disequilibrium rotational distributions might have disappeared by now.

The analogue of this problem is seen in the literature for vibrational energy disposal in nascent *polyatomic* products. Here, one rarely sees population inversion and the distributions are often close to the prior, and often colder (positive λ's). Because the nominal T_{rot} are thus positive (less offensive than negative T's!), they are occasionally quoted, but, as before, for an experiment carried out at fixed E the concept of a Boltzmann temperature is meaningless.

In conclusion, the IT approach to molecular collisions has proved to be a valuable framework upon which to weave the new kinds of detailed experimental (and computer-simulated) state-to-state data in the field of molecular beam and laser chemistry.

Review literature

SHA49, KHI57, JAY63, ASH65, FRI65, KAT67, FAR73, LEV74, LEV74a, BEN75, BER75, BER76, LEV76, LEV76a, LEV76b, BER77, BRO77, LEV77, LEV77a, LEV78, BER79, BER79a, BER79b, LEV79, LEV79a, SMI80, BEN81, JOR81

10 Future directions: laser–molecular beam methods applied to chemical reaction dynamics

10.1. Near-term future developments

It is easy for an author to predict 'progress' in each of the areas to which a chapter has been diligently devoted, but more of a challenge to foresee totally new developments in molecular reaction dynamics without the aid of a crystal ball. So we shall begin in the former mode and end with a few speculations.

There will surely be more emphasis on state-to-state rates and cross sections, requiring improvements in reagent state preparation (e.g., by laser excitation) and product state analysis (e.g., by laser-induced fluorescence or photoionization). Higher resolution (angular and energy) differential cross sections yielding ever more detailed knowledge of potential energy hypersurfaces will be accompanied by progress in *ab initio* methods for computations of these surfaces.

10.2. Orientation dependence and steric factor

Studies of the dependence of reaction probability on the angle of attack as well as the scattering angle (via oriented molecule beam reactions) will provide data against which to test the anisotropic features of the potential surfaces. The dependence of the activation barrier on the relative orientation of the reagents at the distance of closest approach is likely to be significant for virtually all reactions. For example, see the great difference in the barrier to reaction for the collinear versus perpendicular configuration in the reaction of $F + H_2$, based on the best semiempirical calculations, shown in Fig. 10.1.

Analogous to precollision control of mutual orientation of the reagents is postcollision analysis of the polarization of the rotational angular momentum of the products. Polarized laser techniques promise to add a great deal of new knowledge of this kind, which can be used to test the theoretical approaches (both quasiclassical and quantal) to product polarization.

The hitherto elusive steric factor will surely become directly measurable (Chapter 6). As a portent, Fig. 10.2 shows an experimentally derived plot of the probability of reaction (in the 'backscattering' direction in the c.m. system) as a function of the (classically averaged) angle of attack of a Rb atom on a molecule of CH_3I (as depicted on the drawing). Despite extensive experimental averaging ('smearing' over a range of incoming and outgoing angles), it is clear that there is a cone of unfavourable attack angles of at least 45° within which

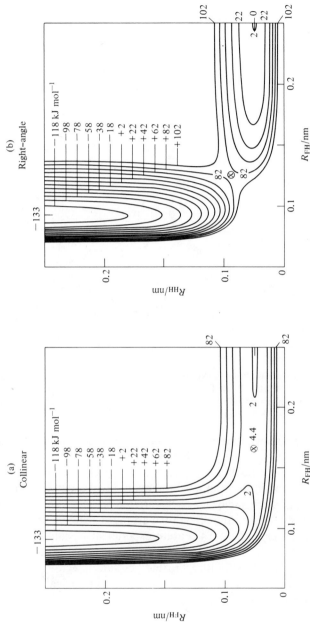

FIG. 10.1. Potential energy surfaces appropriate to describe the $F + H_2$ reaction, representing the best semiempirical knowledge of the FH_2 system. (a) Collinear configuration; (b) perpendicular configuration. The transition state coordinates are designated: for the collinear surface there is a small barrier of height 4.4 kJ mol⁻¹, whereas for the right-angle configuration the barrier is some 85 kJ mol⁻¹. Adapted from J. T. Muckerman in HEN81, Chapter 1; see also ZAR80; details therein.

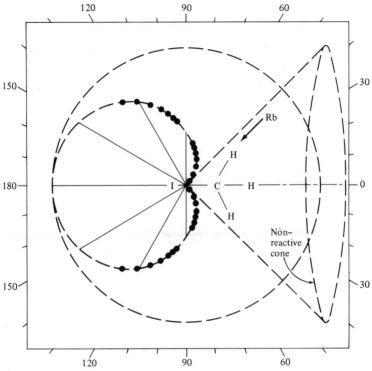

FIG. 10.2. Polar coordinate plot of reaction probability (for c.m. backscattering) as a function of the angle of attack for Rb on CH₃I. The data are from molecular beam measurements using the oriented molecule technique. Adapted from D. H. Parker, K. K. Chakravorty, and R. B. Bernstein, *J. Phys. Chem.* **85**, 466 (1981).

there is a negligible probability of Rb forming RbI product. More specifically, if one identifies backscattering of RbI with head-on collisions (near-zero impact parameters), then only for angles of attack greater than 45° will RbI be formed and ejected in the backward direction. From a detailed analysis of the experiments, the steric factor for backscattered product formation is found to be approximately 0.4, i.e., the orientation-averaged reaction probability (at zero impact parameter) is only 0.4. Future experiments should be capable of evaluating the steric factor for collisions at all impact parameters and thus make contact with the directly measureable absolute integral reaction cross section (and, ultimately, the bimolecular or phenomenological reaction rate coefficient) for unoriented reagents.

10.3. Predissociation and spectroscopy

The use of fast-pulsed molecular beams in conjunction with pulsed tunable lasers will surely be widely adapted for the study of bimolecular reactions as well as the currently fashionable application to unimolecular processes. Laser spectroscopy of ultracold molecules and van der Waals clusters formed in free jet expansions will continue to be an expanding (sic!) field.

Laser-induced predissociation of weakly bound adducts has very interesting theoretical ramifications. Consider the van der Waals molecule Ar $\cdots$ HCl. What will be its lifetime if the HCl bond is laser-excited from its ground to its first vibrational state? There is a large mismatch between the high-frequency H-Cl vibration and the very-low-frequency Ar $\cdots$ M vibration. Will this mismatch decouple or 'protect' the van der Waals bond and thus maintain its integrity? For how long following the HCl bond excitation? Picoseconds? Milliseconds? Theory and experiment should soon be able to cope with questions involving bond-specific excitation and predissociation. Already there are IR laser–molecular beam experiments by G. Scoles et al., W. R. Gentry and coworkers, Y. T. Lee, C. Ng, and others yielding vibrational predissociation spectra for $(NO)_2$, $(NH_3)_2$, $(C_2H_4)_2$, $(HF)_n$, and $(H_2O)_n$ (with $n \lesssim 6$) which yield both dynamic and structural information on the van der Waals molecules.

10.4. Multiphoton processes

Progress in the fields of IR and UV multiphoton excitation (MPE), multiphoton dissociation (MPD), and multiphoton ionization (MPI) of polyatomic molecules has been spectacular. Isotopically selective MPD and MPI have yielded nearly perfect single-stage isotope separations by virtue of the requirement of essentially exact resonance for the absorption of the first photon (which is followed by the absorption of successive photons until the quasicontinuum is reached, beyond which up-pumping is nonselective).

Largely as a result of the laser–molecular beam experiments of Y. T. Lee et al., the interpretation of the IR MPD phenomenon is now almost entirely clarified. The key to the understanding was the determination, under collision-free conditions, of the translational energy distribution of the photofragments as a function of laser intensity and, separately, laser fluence. Without going into detail, it is found that a suitably tailored statistical (RRKM) theory of unimolecular dissociation can explain most of the observations. All the evidence from these experiments suggests that it will be difficult to achieve mode-selective molecular dissociation unless the molecule is such that there is a wide mismatch of normal mode frequencies *and* that an exceedingly fast (ps), intense laser pulse is used.

10.5. Multiphoton ionization and fragmentation

The MPI phenomenon will continue to receive experimental and theoretical attention. Since the new field of MPI—fragmentation (i.e., laser mass spectrometry)—is one with which the author has been concerned (along with co-workers L. Zandee, D. A. Lichtin, S. Datta-Ghosh, K. R. Newton, D. H. Parker, and D. W. Squire), we shall dwell briefly upon the current state of the subject here before projecting the future. Perhaps the simplest case to consider is the resonance-enhanced multiphoton ionization (REMPI) of a simple diatomic molecule, such as NO, at a wavelength of 453 nm (Fig. 10.3). Here the process involves the coherent two-photon excitation of the ground-state molecule to a certain vibrational level of the A state, followed by the absorption

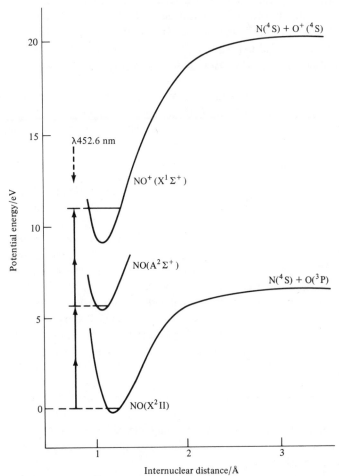

FIG. 10.3. Potential curves for NO and NO$^+$ relevant to the description of the multiphoton ionization process at the indicated laser wavelength. A two-photon excitation to the A state is followed by another two-photon jump to form an excited vibrational state of the NO$^+$ ion. Adapted from L. Zandee and R. B. Bernstein, *J. Chem. Phys.* **71**, 1359 (1979); details and references therein.

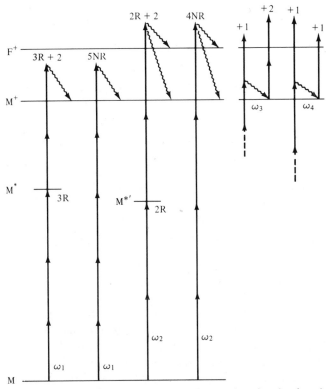

FIG. 10.4. Hypothetical energy level diagrams appropriate for the description of the multiphoton ionization and fragmentation process. Pictured from left to right are the following processes: a three-photon resonant excitation M → M* followed by a two-photon absorption to yield $M^+ + e^-$; a five-photon, direct, nonresonant ionization; a two-photon resonant excitation M → M*′ followed by a two-photon absorption to yield both the parent ion M^+ and the fragment ion F^+; a four-photon equivalent yielding both M^+ and F^+; one 'final' photon leading to M^+ followed by a two-photon absorption by M^+ leading to fragmentation; one 'final' photon leading to M^+ followed by a one-photon absorption by M^+ to form fragment F^+. Adapted from R. B. Bernstein, *J. Phys. Chem.* **86**, (1982); systematics discussed therein.

of two more photons to form a vibrationally hot $NO^+(X^1\Sigma^+)$ state ion, but no fragmentation into atomic ions and/or atoms.

A more complicated process occurs for polyatomic molecules, however. Let us consider the various elementary processes involved in the overall MPI fragmentation of a polyatomic molecule. Figure 10.4 is a simplified energy-level diagram intended to call attention to several of the different possible steps in the mechanism which can be appropriate for different molecules and photon energies.

The sequence can most simply be written

$$M \xrightarrow{n\hbar\omega} M^*$$

$$M^* \xrightarrow{m\hbar\omega} M^+ + e^- \qquad (10.1)$$

$$M^+ \xrightarrow{p\hbar\omega} F^+ + N$$

where M refers to the ground-state neutral molecule, M^* an electronically excited state neutral, M^+ the parent ion, F^+ a fragment ion with its accompanying neutral fragment N. Here n photons are required to excite the molecule, m photons to ionize the M^*, and p photons to fragment the parent ion.

Figure 10.5 is an energy level diagram appropriate for the MPI fragmentation of the methyl iodide molecule. At high laser pulse energies, the observed fragments include most of those energetically allowed, implying the absorption of at least seven near-UV photons.

Figure 10.6 is a schematic drawing of a laser time-of-flight mass spectrometer used to study both resonance-enhanced and nonresonant MPI fragmentation.

Figure 10.7 shows a so-called two-dimensional vibronic/mass spectrum of the caged tertiary amine molecule triethylenediamine, $N(C_2H_4)_3N$, obtained using the laser ionization mass spectrometer of Fig. 10.6. The interpretation of these results and similar data on other polyatomic molecules has been carried out by R. D. Levine and co-workers in terms of a statistical theory somewhat related to the familiar quasiequilibrium theory (QET) of electron-impact mass spectra, due to H. Eyring, H. M. Rosenstock, A. L. Wahrhaftig, and M. B. Wallenstein. Future developments in this field are expected to elucidate the spectroscopy and photofragmentation dynamics of a variety of neutral and ionic polyatomic species. The adaptation of the technique of photo-ion–photoelectron coincidence (PIPECO), as conceived by C. J. Dandy and J. D. Eland, and T. Baer, to the field of MPI fragmentation spectroscopy will prove to be of great interpretive value.

In addition to multiphoton excitation, other methods of laser initiation of chemical change are going to be widely used. One of these is vibrational overtone excitation in the visible and UV regions, as described in Chapter 3. Although the nth overtone absorption strength declines severely with n, it is possible with sufficient photon flux densities to excite very high vibrational overtones and thus selectively pump the molecule to energies that exceed activation barriers to reaction. Possible practical applications abound. For a review of laser-induced chemical processes, including selective photochemistry in the ground electronic state, see LET80.

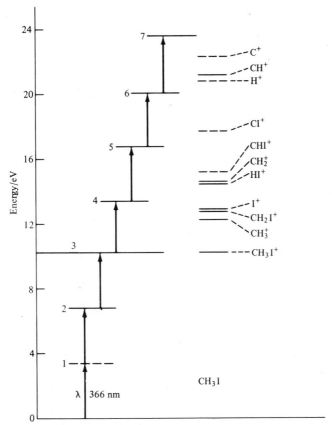

Fig. 10.5. Energy level diagram to account for the multiphoton ionization and fragmentation of CH_3I at the indicated laser wavelength. Adapted from D. H. Parker and R. B. Bernstein, *J. Phys. Chem.* **86**, 60 (1982); details therein.

10.6. Beam-surface interactions

Laser and molecular beam techniques will surely continue to be widely exploited in the ever-growing field of gas-surface interactions. Recent experiments by J. C. Polanyi and R. N. Zare and their co-workers have shown that molecules striking a clean surface can be re-emitted without achieving rotational equilibration with the surface, i.e., the scattered molecules retain a 'memory' of their initial rotational (as well as vibrational) state distributions even after an energetic encounter with the molecules or ions making up the surface. Implications with regard to gas reactions on surfaces and the mechanisms involved in heterogeneous catalysis are significant.

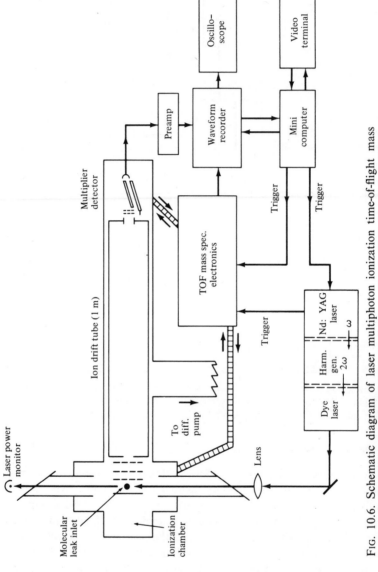

FIG. 10.6. Schematic diagram of laser multiphoton ionization time-of-flight mass spectrometer. Adapted from D. A. Lichtin, S. Datta-Ghosh, K. R. Newton, and R. B. Bernstein, *Chem. Phys. Lett.* **75**, 214 (1980); details therein.

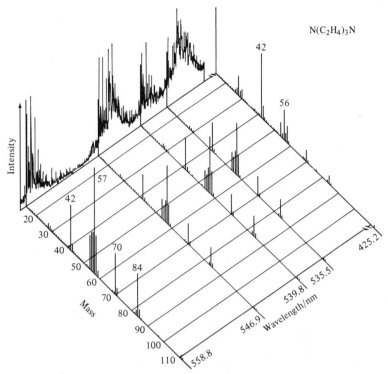

FIG. 10.7. Two-dimensional vibronic/mass spectrum of the caged tertiary amine tri-ethylenediamine, $N(C_2H_4)_3N$, obtained from laser multiphoton ionization TOF mass spectroscopy. *Back panel:* total ionization spectrum; *sticks:* relative intensities of indicated fragment ions. Adapted from reference of Fig. 10.6.

10.7. Laser-augmented collision processes

Finally, we shall devote the remainder of this chapter to a consideration of a new area of laser-augmented collisions and reactions. There is good reason to believe that strong laser fields can influence the transition state and possibly thereby alter the course of an elementary chemical reaction, as pointed out by A. M. Lau, by T. F. George et al., and by W. H. Miller and co-workers and experimentally elucidated by S. Harris et al. and by J. Weiner and others. Figure 10.8 is a diagrammatic sketch distinguishing the new application of lasers to chemical dynamics: not *pre*collision, not *post*collision, but rather '*in*-collision' excitation to effect a perturbation of the transition state. Figure 10.9 is a (hypothetical) set of potential energy curves (cuts through the hypersurface) appropriate to the collisional ionization process:

$$A + BC \rightarrow A^+ + BC^- \tag{10.2}$$

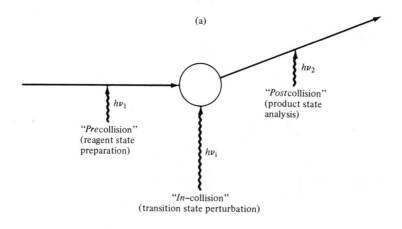

(a)

$h\nu_2$

"*Post*collision"
(product state
analysis)

$h\nu_1$

"*Pre*collision"
(reagent state
preparation)

$h\nu_i$

"*In*-collision"
(transition state perturbation)

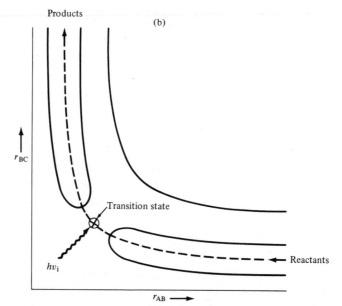

Products

(b)

r_{BC}

Transition state

$h\nu_i$

Reactants

$r_{AB} \longrightarrow$

FIG. 10.8. Schematic illustration to assist in interpretation of experiments on the 'in-collision' excitation of an elementary reaction. (a): Scheme; (b) potential surface, indicating irradiation of transition state to influence the course of reaction. See T. F. George et al., GEO77, and text for further details.

In the absence of the laser field, the situation is as described in Chapter 8. When the diabats are of the same symmetry, the noncrossing rule implies the splitting into two adiabats, and the barrier to the formation of [ABC] and thus any rearrangement reaction is E_b. However, in the presence of intense radiation fields, i.e., upon irradiation by a high-power laser, the splitting ΔE may, under certain conditions, increase and thereby lower the barrier to the value E_b^*, thus permitting the rearrangement reaction to occur at lower collision energy. Thus the laser might serve to reduce the activation barrier to chemical reaction. The magnitude of the dependence of the splitting of the adiabats upon the laser field has been estimated, and it appears that with currently available laser powers some systems can be found which would be susceptible to this laser augmentation of reactivity (via perturbation of transition state).

A simpler but somewhat related experiment has already been carried out to demonstrate in-flight laser excitation leading to Penning ionization. The process under investigation, by J. Weiner et al., is the reaction

$$Kr^*(^3P_2) + Xe(^1S_0) \rightarrow Kr(^1S_0) + Xe^+ + e^- \qquad (10.3)$$

endoergic by 2.21 eV. Thus at thermal collision energies, Penning ionization is energetically forbidden. However, upon irradiation with a laser tuned in the neighborhood of this difference energy ($\lambda \approx 561$ nm), Penning ionization (i.e., Xe^+ formation) was indeed observed, indicating that the in-collision Kr^*–Xe

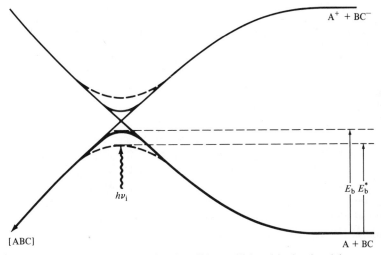

FIG. 10.9. Hypothetical potentials for describing collisional ionization (charge transfer) of A + BC, in which the laser irradiation may serve to lower the barrier (from E_b to E_b^*, as shown) and thus make *less* probable the charge transfer reaction, according to an LZ model, while allowing the formation of ABC with higher probability. See text and reference of Fig. 10.8 for details.

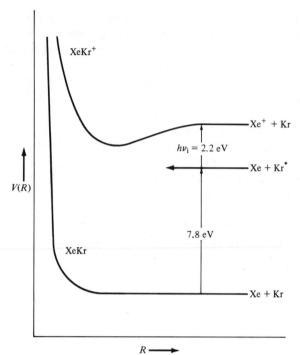

FIG. 10.10. Potential curves for the Xe–Kr system to elucidate the laser-induced Penning ionization process, Kr* + Xe → Kr + Xe$^+$ + e$^-$. Penning ionization is energetically forbidden but can be induced by tuning the laser to a wavelength of 561 nm (ΔE = 2.2 eV), i.e., in-flight, collision-assisted laser ionization. Adapted from J. Weiner, *J. Chem. Phys.* **72**, 2856 (1980). See also WEI79 and GEO77 for further details.

pair, i.e., the transitory Kr* $\cdots$ Xe molecule, absorbs the 2.2-eV photon to yield Kr + Xe* (Fig. 10.10).

This process is amenable to theoretical treatment and appears to be explainable semiquantitatively. This is similar to the case of collision-induced absorption in the IR, e.g., absorption in the course of 'fly-by collisions' of dissimilar (e.g., He and Ar) atoms, as studied experimentally by H. L. Welsh and theoretically by H. B. Levine, G. Birnbaum, D. A. McQuarrie, R. B. Bernstein, and others. Here, one requires a knowledge of both the interaction potential and the key matrix elements to carry out a full analysis and computer-simulation of the experimental results.

10.8. Example of a laser-assisted reaction

Returning to the more chemically interesting subject of laser-assisted reactions, it does appear that lasers can provide radiation fields of sufficient intensity to influence intermolecular interactions during the subpicosecond interval during which the system is neither 'reagents' nor 'products,' but rather 'transition state' or 'activated complex.' Although true laser catalysis, in this sense, has not yet been definitively observed, an experiment with nearly this attribute has been carried out. P. R. Brooks, R. F. Curl et al. reported that in a crossed molecular beam–laser experiment involving K + $HgBr_2$ they observed light absorption by the collision complex. Figure 10.11 characterises the experiment.

Figure 10.11a is the relevant energy level diagram, and Fig. 10.11b is a schematic of the experimental arrangement. An atomic beam of K is crossed by a molecular beam of $HgBr_2$. It is known from previous work that the ground-state reaction is exoergic by 0.86 eV and proceeds by way of a long-lived complex as follows:

$$K + BrHgBr \rightleftarrows [KBrHgBr] \rightarrow KBr + HgBr(X^2\Sigma^+) \qquad (10.4)$$

The new idea was to irradiate the beam intersection volume with a laser at a wavelength (595 nm) that can excite neither the reagents nor the products, but might affect the transition state. Using an intense flashlamp–pumped dye laser with the target volume intracavity, they observed weak chemiluminescence at 500 nm (Fig. 10.11a). They attributed this fluorescence to emission from electronically excited $HgBr^*(B^2\Sigma)$. If so, the laser has served to excite the activated complex. The following sequence can be suggested to explain the results:

$$K + BrHgBr \rightleftarrows [KBrHgBr] \qquad \text{complex formation} \qquad (10.5a)$$

$$[KBrHgBr] + h\nu(\lambda 595 \text{ nm}) \rightarrow$$

$$\qquad KBr + HgBr^*(B^2\Sigma) \qquad \text{photodecomposition} \qquad (10.5b)$$

$$HgBr^*(B^2\Sigma) \rightarrow HgBr(X^2\Sigma^+) + h\nu(\lambda 500 \text{ nm}) \qquad \text{fluorescence} \qquad (10.5c)$$

Unfortunately, there are two difficulties with the above scheme. First, since the energy of the complex is bound to be below that of the reagents, a laser wavelength *shorter* than 595 nm would be required to bring the complex to the indicated level for KBr + $HgBr^*(B)$ such that the emitted photon energy would coincide with the B-X excitation energy of HgBr. (However, collision energy could play a role in ameliorating this situation.) Second, the possibility of multiphoton effects due to the high laser pulse power levels cannot yet be ruled out.

Nevertheless, such studies of the influence of the absorption of laser radiation upon the transition state (and related experiments on the emission of radiation *by* the transition state, such as those of J. C. Polanyi and co-workers P. Arrowsmith, F. E. Bartoszek, S. H. Bly, T. Carrington, and P. E. Charters) will be extremely valuable, first as a probe of the structure of the activated complex but later as a tool to influence the very course of an elementary chemical reaction.

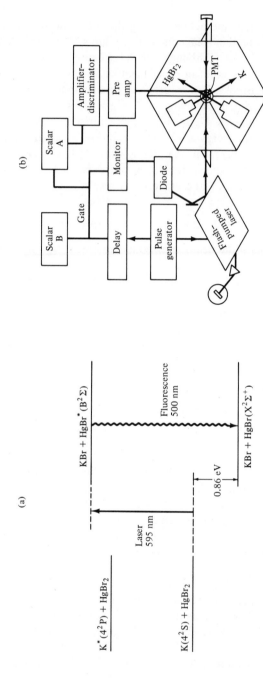

FIG. 10.11. Energy level diagram intended to elucidate the laser-assisted chemistry experiment on K + HgBr$_2$ by P. Hering, P. R. Brooks, R. F. Curl, R. S. Judson, and R. S. Lowe, *Phys. Rev. Lett.* **44**, 687 (1980). The ground-state reaction to yield KBr + HgBr is exoergic and is known to be fast. (a) Laser radiation at 595 nm may excite the collision complex and produce KBr + HgBr*; the latter product would then fluoresce at 500 nm as shown. The excited state of K (shown) should be of no relevance to the experiment. (b) A schematic view of the apparatus, consisting of the two oven sources providing reagent beams at 120° incidence and the reaction occurring intracavity of the laser, with fluorescence being monitored by a photomultiplier out of the plane. See P. Hering et al., reference above, for details.

Perhaps, then, it is not merely a speculation but rather a straightforward extrapolation to conclude that molecular beam and laser techniques will lead to truly dramatic developments in the field of chemical reaction dynamics in the years to come. A more significant question, however, remains to be answered: will all of the above ever really influence 'real-world' chemistry? (The author places his bet: YES!)

Review literature

BYL77, GEO77, GEO77a, GRU78, SHA78, BES79, BOE79, GLO79, KOM79a, LAU79, ROC79, JOH80, LEE80, LET80, PHY80, ZAR80, ZEW80, BEN81, CAN81, JOH81, JOR81, RON81

Bibliography of review literature

AMD66   Amdur, I., and J. D. Jordan in ROS66, p. 29.

AND65   Anderson, J. B., R. P. Andres, and J. B. Fenn, *Adv. At. Mol. Phys.* **1**, 345 (1965).

AND66   ——, ——, ——, *Adv. Chem. Phys.* **10**, 275 (1966).

ASH65   Ash, R., *Information Theory.* New York: Wiley (Interscience), 1965.

BAE75   Baede, A. P., *Adv. Chem. Phys.* **30**, 463 (1975).

BAG79   Bagratashoili, V. N., V. S. Doljikov, V. S. Letokhov, and E. A. Ryabov in KOM79a, p. 179.

BAL75   Balint-Kurti, G. G., *Adv. Chem. Phys.* **30**, 137 (1975).

BAL75a   —— in *International Review of Science MTP, Physical Chemistry,* Series II (A. D. Buckingham and C. A. Coulson, eds.), p. 283. London: Butterworths, 1975.

BAL77   —— and R. N. Yardley in FAR77, p. 77.

BAR81   Baronavski, A., M. E. Umstead, and M. C. Lin in JOR81, Part 2, p. 85.

BAU78   Bauer, S. H., *Chem. Rev.* **78**, 147 (1978).

BEN75   Ben-Shaul, A., and G. L. Hofacker in *Handbook of Chemical Lasers* (J. F. Bott and R. W. Gross, eds.), p. 579. New York: Wiley, 1975.

BEN81   ——, Y. Haas, K. K. Kompa, and R. D. Levine, *Laser and Chemical Change,* Springer Series in Chemical Physics, Vol. 10. Berlin: Springer-Verlag, 1981.

BER64   Bernstein, R. B., *Science* **144**, 141 (1964).

BER64a   —— in MCD64, p. 895.

BER66   ——, *Adv. Chem. Phys.* **10**, 75 (1966).

BER67   —— and J. T. Muckerman, *Adv. Chem. Phys.* **12**, 389 (1967).

BER67a   —— in 'Molecular Forces,' *Pontif. Acad. Sci. Scr. Varia* **31**, 25 (1967).

BER70   Berry, R. S., in SCH70, p. 193.

BER70a   —— in SCH70, p. 229.

BER71   Bernstein, R. B., *Isr. J. Chem.* **9**, 615 (1971).

BER73   —— and A. M. Rulis in FAR73, p. 293.

BER73a   ——, *Comments At. Mol. Phys.* **D4**, 43 (1973).

BER75   —— and R. D. Levine, *Adv. At. Mol. Phys.* **11**, 215 (1975).

BER75a   ——, *Isr. J. Chem.* **14**, 79 (1975).

BER75x   Berry, M. J., *Ann. Rev. Phys. Chem.* **26**, 259 (1975).

BER76   Bernstein, R. B., *Intl. J. Quant. Chem.* **S10**, 267 (1976).

BER77   —— in BRO77, p. 3.

BER78   —— in ZEW78, p. 384.

BER78a   ——, *Pure Appl. Chem.* **50**, 781 (1978).

BER79   ——, ed. *Atom-Molecule Collision Theory—A Guide for the Experimentalist.* New York: Plenum, 1979.

BER79a   ——, *Adv. Atom. Mol. Phys.* **15**, 167 (1979).

BER79b   —— in GLO79, p. 161.

BES79   Beswick, J. A., and J. Durup in GLO79, p. 385.

BIR75   Birely, J. H., and J. L. Lyman, *J. Photochem.* **4**, 269 (1975).

BLO79   Bloembergen, N., and E. Yablonovitch in KOM79a, p. 117.

BOE79   Boesl, U., W. J. Neusser, and E. W. Schlag in KOM79a, p. 219.

BRO76   Brooks, P. R., *Science* **193**, 11 (1976).

BRO77   —— and E. F. Hayes, eds., *State-to-State Chemistry*, A.C.S. Symposium Ser. No. 56. Washington, D.C.: American Chemical Society, 1977.

BUC75   Buck, U., *Adv. Chem. Phys.* **30**, 313 (1975).

BUN70   Bunker, D. L., in SCH70, p. 355.

BUN73   —— and E. A. Goring-Simpson in FAR73, p. 521.

BYL77   Bylinsky, G., 'Laser Alchemy Is Just Around the Corner,' *Fortune*, September 1977, p. 186.

CAN81   Cantrell, C. D., ed., *Multiple Photon Excitation and Dissociation of Polyatomic Molecules*, Topics in Current Physics. Berlin: Springer-Verlag, 1981.

CAR72   Carrington, T., and J. C. Polanyi in *Physical Chemistry, M.T.P. International Review of Science* (J. C. Polanyi, ed.), Ser. 1, Vol. 9, p. 135. London: Butterworths, 1972.

CHI74   Child, M. S., *Molecular Collision Theory.* New York: Academic Press, 1974.

CHI76   —— in MIL76, Part B, p. 171.

CHI79   —— in BER79, p. 427.

COH76   Cohen, N., and J. F. Bott in GRO76, p. 33.

CRO80   Crosley, D. R., ed., *Laser Probes for Combustion Chemistry*, A.C.S. Symposium Series 134. Washington, D.C.: American Chemical Society 1980.

CRU73   Cruse, H. W., P. J. Dagdigian, and R. N. Zare in FAR73, p. 277.

DAV79   Davidovits, P., and D. L. McFadden, *The Alkali Halide Vapors.* New York: Academic Press, 1979.

DEH66   DeHeer, F. J., *Adv. At. Mol. Phys.* **2**, 328 (1966).

DEM71   Demtroder, W., *Laser Spectroscopy*, Topics in Current Chemistry, Vol. 17. Berlin: Springer, 1971.

DEM77   ——, *Principles and Techniques of Laser Spectroscopy.* Berlin: Springer, 1977.

DIE79   Diestler, D. J., in BER79, p. 655.

DIN73   Ding, A. M., L. J. Kirsch, D. S. Perry, J. C. Polanyi, and J. L. Schreiber in FAR73, p. 252.

DUB72   Dubrin, J., and M. J. Henchman in *Physical Chemistry M.T.P. International Review of Science* (J. C. Polanyi, ed.), Ser. 1, Vol. 9, p. 213. London: Butterworths, 1972.

DUB73   ——, *Ann. Rev. Phys. Chem.* **24**, 97 (1973).

EWI77   Ewing, J. J., in MOO77, Vol. II, p. 241.

EYA81   Eyal, M., V. Agam, and F. R. Grabiner in JOR81, Part 2, p. 43.

EYR74   Eyring, H., W. Jost, and D. Henderson, eds., *Physical Chemistry—An Advanced Treatise*, Vol. VIA. New York: Academic Press, 1974.

EYR75   Vol. VIB of EYR74.

EYR80   Eyring, H., S. H. Lin, and S. M. Lin, *Basic Chemical Kinetics*. New York: Wiley-Interscience, 1980.

FAR73   Faraday Discussion of the Chemical Society, *Molecular Beam Scattering*, Vol. 55 (1973).

FAR74   Farrar, J. M., and Y. T. Lee, *Ann. Rev. Phys. Chem.* **25**, 357 (1974).

FAR77   Faraday Discussion of the Chemical Society, *Potential Energy Surfaces*, Vol. 62 (1977).

FAR79   ——, *Kinetics of State-Selected Species*, Vol. 67 (1979).

FAU77   Faubel, M., and P. J. Toennies, *Adv. At. Mol. Phys.* **13**, 229 (1977).

FEL71   Feld, M. S., and N. Kurnit, eds., *Fundamental and Applied Laser Physics*. New York: Wiley, 1973.

FIT63   Fite, W. L., and S. Datz, *Ann. Rev. Phys. Chem.* **14**, 61 (1963).

FLU73   Fluendy, M. A. D., and K. P. Lawley, *Chemical Applications of Molecular Beam Scattering*. London: Chapman and Hall, 1973.

FLY74   Flynn, G. E., P. Pechukas, R. C. Stern, and R. N. Zare, eds., N.S.F. Workshop Report, *Gas-Phase Molecular Interactions and the Nation's Energy Problem*, Columbia University. Arden, N.Y.: Harriman House, 1974.

FOR59   Ford, K. W., and J. A. Wheeler, *Ann. Phys. (N.Y.)* **7**, 259 (1959).

FOR73   Forst, W., *Theory of Unimolecular Reactions*. New York: Academic Press, 1973.

FRI65   Fristrom, R. L., and A. A. Westenberg, *Flame Structure*. New York: McGraw-Hill, 1965.

GEI77   Geis, M. W., H. Dispert, T. L. Budzynski, and P. R. Brooks in BRO77, p. 102.

GEO73   George, T. F., and J. Ross, *Ann. Rev. Phys. Chem.* **24**, 263 (1973).

GEO77   ——, I. H. Zimmermann, J. M. Yuan, J. R. Laing, and P. L. DeVries, *Acc. Chem. Res.* **10**, 449 (1977).

GEO77a   ——, ed., *Theoretical Aspects of Laser Radiation and Its Interaction with Atomic and Molecular Systems*. Rochester, N.Y.: N.S.F., University of Rochester, 1977.

GLA41   Glasstone, S., K. J. Laidler, and H. Eyring, *Theory of Rate Processes*. New York: McGraw-Hill, 1941.

GLO79   Glorieux, P., D. Lecler, and R. E. Vetter, eds., *Chemical Photophysics.* CNRS Paris: Les Houches, 1979.

GOL64   Goldberger, M. L., and K. M. Watson, *Collision Theory.* New York: Wiley, 1964.

GOR68   Gordon, R. G., W. Klemperer, and J. I. Steinfeld, *Ann. Rev. Phys. Chem.* **19**, 215 (1968).

GOR73   —— in FAR73, p. 22.

GRA77   Grant, E. R., M. J. Coggiola, Y. T. Lee, P. A. Schulz, A. S. Sudbo, and Y. R. Shen in BRO77, p. 72.

GRE66   Greene, E. F., A. L. Moursund, and J. Ross, *Adv. Chem. Phys.* **10**, 135 (1966).

GRI75   Grice, R., *Adv. Chem. Phys.* **30**, 247 (1975).

GRI79   —— in FAR79, p. 16.

GRO76   Gross, R. W., and J. F. Bott, eds., *Handbook of Chemical Lasers.* New York: Wiley, 1976.

GRU78   Grunwald, E., D. F. Dever, and P. M. Keehn, *Megawatt Infrared Laser Chemistry.* New York: Wiley, 1978.

HAA81   Haas, Y., in JOR81, Part 1, p. 713.

HAL81   Hall, R. B., A. Kaldor, D. M. Cox, J. A. Horsley, P. Rabinowitz, G. M. Kramer, R. G. Bray, and E. T. Maas in JOR81, Part 1, p. 639.

HAR68   Hartmann, H., ed., *Chemische Elementarprozesse.* Berlin: Springer-Verlag, 1968.

HAR77   Harper, P. G., and B. S. Wherret, eds., *Nonlinear Optics.* New York: Academic Press, 1977.

HAS64   Hasted, J. B., *Physics of Atomic Collisions.* New York: Elsevier, 1964, 1972.

HAS76   Hase, W. L., in MIL76, Part B, p. 121.

HAS81   —— in TRU81, p. 1.

HEN81   Henderson, D., ed., *Theoretical Chemistry.* Vol. 6: *Theory of Scattering: Papers Dedicated to Henry Eyring.* New York: Academic Press, 1981.

HER62   Herschbach, D. R., *Disc. Far. Soc.* **33**, 149 (1962).

HER65   ——, *Appl. Opt. Suppl.* **2**, 128 (1965).

HER66   ——, *Adv. Chem. Phys.* **10**, 319 (1966).

HER73   —— in FAR73, p. 233.

HER73a   —— in FAR73, p. 303.

HER76   ——, *Pure Appl. Chem.* **47**, 61 (1976).

HER77   —— in FAR77, p. 447.

HER79   Herm, R. R., in DAV79, p. 189.

HIN76   Hinkley, E. D., K. W. Nill, and F. A. Blum, *Laser Spectroscopy of Atoms and Molecules* (H. Walther, ed.), Chap. 2. Berlin: Springer-Verlag, 1976.

HIN76a   ——, ed., *Laser Monitoring of the Atmosphere.* Berlin: Springer, 1976.

HIR54   Hirschfelder, J. O., C. F. Curtiss, and R. B. Bird, *Molecular Theory of Gases and Liquids.* New York: Wiley, 1954, 1964.

HIR67   ——, ed., *Intermolecular Forces,* Adv. Chem. Phys. **12**, (1967).

HOU81   Houston, P. L., in JOR81, Part 1, p. 625.

JAC76   Jacobs, S. F., M. Sargent, M. O. Scully, and C. T. Walker, *Laser Photochemistry, Tunable Lasers and Other Topics.* Reading, Mass.: Addison-Wesley, 1976.

JAY63   Jaynes, E. T., *Statistical Physics,* Brandeis Lectures, Vol. 3, p. 81. New York: Benjamin, 1963.

JOH80   Johnson, P. M., *Acc. Chem. Res.* **13**, 20 (1980).

JOH81   —— and C. E. Otis, *Ann. Rev. Phys. Chem.* **32**, 139 (1981).

JOR81   Jortner, J., R. D. Levine, and S. A. Rice, eds., *Photoselective Chemistry,* Parts 1 and 2, Adv. Chem. Phys. **47** (1981).

JOR81a   —— and R. D. Levine in JOR81, Part 1, p. 1.

KAR70   Karplus, M., in SCH70, pp. 320, 372.

KAT67   Katz, A., *Principles of Statistical Mechanics: The Information Theory Approach.* San Francisco: Freeman, 1967.

KEM75   Kempter, V., *Adv. Chem. Phys.* **30**, 417 (1975).

KHI57   Khinchin, A. K., *Mathematical Foundations of Information Theory.* New York: Dover, 1957.

KIN72   Kinsey, J. L., in *Physical Chemistry, M.T.P. International Review of Science* (J. C. Polanyi, ed.), Ser. 1, Vol. 9, p. 173. London: Butterworths, 1972.

KIN77   ——, *Ann. Rev. Phys. Chem.* **28**, 349 (1977).

KNE80   Kneba, M., and J. Wolfrum, *Ann. Rev. Phys. Chem.* **31**, 47 (1980).

KOM73   Kompa, K. L., *Chemical Lasers,* Topics in Current Chemistry, Vol. 37. Berlin: Springer, 1973.

KOM73a   ——, *Top. Curr. Chem.* **37**, 1 (1973).

KOM79   —— and H. Walther, eds., *High-Power Lasers and Applications.* Berlin: Springer, 1979.

KOM79a   —— and S. D. Smith, eds., *Laser-Induced Processes in Molecules,* Springer Series in Chemical Physics, Vol. 6. Berlin: Springer, 1979.

KUN76   Kuntz, P. J., in MIL76, Part B, p. 53.

KUN79   —— in BER79, p. 79.

KUP79   Kuppermann, A., in GLO79, p. 293.

KUP81   —— in HEN81, p. 80.

KWE79   Kwei, G. H., in DAV79, p. 441.

LAU79   Lau, A. M. F., in KOM79a, p. 163.

LAW75   Lawley, L. P., ed., *Molecular Scattering: Physical and Chemical Applications,* Adv. Chem. Phys. **30** (1975).

LEE72   Lee, Y. T., in *Physics of Electronic and Atomic Collisions* (T. R. Govers and F. J. DeHeer, eds.), p. 357. Amsterdam: North Holland, 1972.

LEE80   —— and Y. R. Shen, *Phys. Today* **33**(11), 52 (1980).

LEM77   Lemont, S., and G. W. Flynn, *Ann. Rev. Phys. Chem.* **28**, 261 (1977).

LEN74   Lengyel, B. A., *Lasers,* 2nd ed. New York: Wiley, 1974.

LES70   Lester, W. A. Jr., ed., *Potential Energy Surfaces in Chemistry.* San Jose, Calif.: IBM, 1970.

LES76   —— in MIL76, p. 1.

LET77   Letokhov, V. S., and V. P. Chebotayev, *Nonlinear Laser Spectroscopy.* Berlin: Springer-Verlag, 1977.

LET80   ——, *Phys. Today* **33**(11), 34 (1980).

LEV69   Levine, R. D., Quantum Mechanics of Molecular Rate Process. Oxford: Clarendon Press, 1969.

LEV72   —— in *Theoretical Chemistry, M.T.P. International Review of Science* (W. B. Brown, ed.), Ser. 1, Vol. 1, p. 229. London: Butterworths, 1972.

LEV73   —— and R. B. Bernstein in FAR73, p. 100.

LEV74   —— and ——, *Molecular Reaction Dynamics.* Oxford: Clarendon Press, 1974; also designated MRD.

LEV74a   —— and ——, *Acc. Chem. Res.* **7**, 393 (1974).

LEV76   —— and —— in MIL76, Part B, p. 323.

LEV76a   —— in *The New World of Quantum Chemistry* (B. Pullman and R. W. Parr, eds.). Boston: Reidel, 1976.

LEV76b   —— and J. Jortner, eds., *Molecular Energy Transfer.* New York: Halsted, Wiley, 1976.

LEV77   —— in BRO77, p. 226.

LEV77a   —— and A. Ben-Shaul in *Chemical and Biochemical Applications of Lasers* (C. B. Moore, ed.), Vol. 2, p. 145. New York: Academic Press, 1977.

LEV77x   Levy, D. H., L. Wharton, and R. E. Smalley in MOO77, Vol. II, p. 1.

LEV78   Levine, R. D., *Ann. Rev. Phys. Chem.* **29**, 59 (1978).

LEV79   —— and J. L. Kinsey in BER79, p. 693.

LEV79a   —— in *Maximum Entropy Formalism* (R. D. Levine and M. Tribus, eds.). Cambridge, Mass.: MIT Press, 1979.

LEV79x   Levy, M. R., in *Progress in Reaction Kinetics,* Vol. 10, p. 1. Oxford: Pergamon Press, 1979.

LIG71   Light, J. C., *Adv. Chem. Phys.* **19**, 1 (1971).

LIG79   —— in BER79, p. 467.

LOS79   Los, J., and A. W. Kleyn in DAV79, p. 275.

MAI69   Maitland, A., and M. H. Dunn, *Laser Physics.* Amsterdam: North-Holland, 1969.

MAI81   ——, M. Rigby, E. B. Smith, and W. A. Wakeham, *Intermolecular Forces: Their Origin and Determination.* Oxford: Clarendon Press, 1981.

MAP72   Mapleton, R. A., *Theory of Charge Exchange.* New York: Wiley (Interscience), 1972.

MAS71   Massey, H. S. W., in *Electronic and Ionic Impact Phenomena* (H. S. W. Massey, E. H. S. Burhop, and H. B. Gilbody, eds.), 2nd ed., Vol. 3, *Slow Collisions of Heavy Particles*. Oxford: Clarendon Press, 1971.

MCC72   McClure, G. A., and J. M. Peek, *Dissociation in Heavy Particle Collisions*. New York: Wiley (Interscience), 1972.

MCD64   McDaniel, E. W., *Collision Phenomena in Ionized Gases*. New York: Wiley, 1964.

MCD64x   McDowell, M. R. C., ed., *Atomic Collision Processes*. Amsterdam: North-Holland, 1964.

MCG75   McGowan, J. W., ed., *The Excited State in Chemical Physics, Part 1*, Adv. Chem. Phys., **28** (1975).

MCG81   ——, ed., *The Excited State in Chemical Physics, Part 2*, Vol. 45 of Adv. Chem. Phys. **45** (1981).

MIC75   Micha, D. A., *Adv. Chem. Phys.* **30**, 7 (1975).

MIC81   —— in TRU81, p. 685.

MIL67   Miller, W. B., S. A. Safron, and D. R. Herschbach in FAR67, p. 108.

MIL76   Miller, W. H. , ed., *Dynamics of Molecular Collisions*, Parts A and B. New York: Plenum, 1976.

MOO74   Moore, C. B., ed., *Chemical and Biological Applications of Lasers*, Vol. I. New York: Academic Press, 1974.

MOO76   Mooradian, A., T. Jaeger, and P. Stokseth, eds., *Tunable Lasers and Applications*. Berlin: Springer, 1976.

MOO77   Moore, C. B., ed., *Chemical and Biological Applications of Lasers*, Vols. II, III. New York: Academic Press, 1977.

MOO79   ——, *Chemical and Biological Applications of Lasers*, Vol. IV. New York: Academic Press, 1979.

MOO79a   —— and I. W. M. Smith in FAR79, p. 146.

MOO81   Moore, J. W., and R. G. Pearson, *Kinetics and Mechanism*. New York: Wiley-Interscience, 1981.

MOT33   Mott, N. F., and H. S. W. Massey, *The Theory of Atomic Collisions*, Oxford: Clarendon Press, 1933, 1949, 1965.

MUC81   Muckerman, J. T., in HEN81, p. 1.

MUS66   Muschlitz, E. E. Jr., *Adv. Chem. Phys.* **10**, 171 (1966).

NIK74   Nikitin, E. E., *Theory of Elementary Atomic and Molecular Processes in Gases*. Oxford: Clarendon Press, 1974.

NIK74a   —— in EYR74, p. 187.

OKA78   Okabe, H., *Photochemistry of Small Molecules*. New York: Wiley, 1978.

PAO77   Pao, Y. H., ed., *Optoacoustic Spectroscopy and Detection*. New York: Academic Press, 1977.

PAR73   Parson, J. M., K. Shobatake, Y. T. Lee, and S. A. Rice in FAR73, p. 344.

PAU65   Pauly, H., and J. P. Toennies, *Adv. At. Mol. Phys.* **1**, 201 (1965).

PAU68   —— and ——, *Methods Exptl. Phys.* **7A**, 261 (1968).

PAU73   —— in FAR73, p. 191.

PAU75   —— in EYR75, Vol. VIB, p. 553.

PAU79   —— in BER79, p. 111.

PHY80   *Physics Today,* 'Laser Chemistry,' **33**(11), 25 (1980).

PIM76   Pimentel, G. C., and K. L. Kompa in GRO76, p. 1.

POL72   Polanyi, J. C., *Acc. Chem. Res.* **5**, 161 (1972).

POL73   —— in FAR73, p. 389.

POL74   —— and J. L. Schreiber in EYR74, p. 383.

POL77   —— and —— in FAR77, p. 267.

POR74   Porter, R. N., *Ann. Rev. Phys. Chem.* **25**, 317 (1974).

POR76   —— and L. M. Raff in MIL76, Part B, p. 1.

RAM56   Ramsey, N. F., *Molecular Beams.* Oxford: Clarendon Press, 1956.

RED78   Reddy, K. V., R. G. Bray, and M. J. Berry in ZEW78, p. 48.

RED79   —— and M. J. Berry in FAR79, p. 222.

REI81   Reisler, A., and C. Wittig in JOR81, Part 1, p. 679.

REU75   Reuss, J., *Adv. Chem. Phys.* **30**, 389 (1975).

RHO79   Rhodes, C. K., ed., *Excimer Lasers,* Topics in Applied Physics, Vol. 30. Berlin: Springer, 1979.

ROB72   Robinson, P. J., and K. A. Holbrook, *Unimolecular Reactions.* New York: Wiley-Interscience, 1972.

ROC79   Rockwood, S. D., in KOM79a, p. 3.

RON81   Ronn, A. M., in JOR81, Part 1, p. 661.

ROS62   Ross, J., and E. F. Greene in *Energy Transfer in Gases,* Solvay Institute 12th Chemistry Conference, p. 363. New York: Interscience, 1962.

ROS65   ——, J. C. Light, and K. E. Shuler in *Kinetic Processes in Gases and Plasmas* (A. R. Hochstim, ed.), p. 281. New York: Academic Press, 1965.

ROS66   ——, ed., *Molecular Beams,* Adv. Chem. Phys. **10** (1966). New York: Wiley, 1966.

ROS69   ——, J. C. Light, and K. E. Shuler in *Kinetic Processes in Gases and Plasmas* (A. R. Hochstim, ed.), p. 281. New York: Academic Press, 1969.

ROS70   —— and E. F. Greene in SCH70, p. 86.

SAR74   Sargent, M., M. O. Scully, and W. E. Lamb Jr., *Laser Physics.* Reading, Mass.: Addison-Wesley, 1974.

SCH69   Schlier, C., *Ann. Rev. Phys. Chem.* **20**, 191 (1969).

SCH70   —— ed., *Molecular Beams and Reaction Kinetics.* New York: Academic Press, 1970.

SCH72   Schaefer, III, H. F., *Electronic Structure of Atoms and Molecules.* London: Addison-Wesley, 1972.

SCH77   Schafer, F. P., ed., *Dye Lasers.* Topics in Applied Physics, Vol. 1, 2nd ed. Berlin: Springer, 1977.

SCH77x   Schmatjko, K. J., and J. Wolfrum in 16th Intl. Symp. on Combustion. Pittsburgh: The Combustion Institute, 1977.

SCH79   Schaefer, III, H. F., in BER79, p. 45.

SCH81   Schatz, G. C., in TRU81, p. 287.

SEC74   Secrest, D., Ann. Rev. Phys. Chem. 24, 379 (1974).

SHA49   Shannon, C. E., and W. Weaver, Mathematical Theory of Communication. Urbana: University of Illinois Press, 1949.

SHA78   Shank, C. V., E. P. Ippen, and S. L. Shapiro, eds., Picosecond Phenomena, Springer Series in Chemistry and Physics, Vol. 4. Berlin: Springer, 1978.

SHE77   Shen, Y. R., ed., Nonlinear Infrared Generation, Topics in Applied Physics, Vol. 16. Berlin: Springer, 1977.

SIE71   Siegman, A. E., An Introduction to Lasers and Masers. New York: McGraw-Hill, 1971.

SMI72   Smith, K., The Calculation of Atomic Collision Processes. New York: Wiley-Interscience, 1972.

SMI79   Smith, I. W. M., ed., Physical Chemistry of Fast Reactions II. Reactions Dynamics, p. 1. New York: Plenum, 1979.

SMI80   ——, Kinetics and Dynamics of Elementary Gas Reactions. London: Butterworths, 1980.

STE66   Stebbings, R. F., Adv. Chem. Phys. 10, 195 (1966).

STE68   ——, Adv. At. Mol. Phys. 4, 299 (1968).

STE74   Steinfeld, J. I., Molecules and Radiation. New York: Harper and Row, 1974.

STE75   ——, ed., Electronic Transition Lasers. Cambridge, Mass.: MIT Press, 1975.

STE76   —— and M. Wrighton, eds., The Laser Revolution in Energy-Related Chemistry. Cambridge, Mass.: MIT Press, 1976.

STE78   ——, ed., Laser and Coherence Spectroscopy. New York: Plenum, 1981.

STE81   ——, Laser-Induced Chemical Processes. New York: Plenum, 1981.

STI79   Stitch, M. L., ed., Laser Handbook, Vol. IIIb. Amsterdam: North Holland, 1979.

STO79   Stolte, S., and J. Reuss in BER79, p. 201.

SUD79   Sudbo, A. S., P. A. Schulz, D. J. Krajnovich, Y. R. Shen, and Y. T. Lee in GLO79, p. 461.

SVE76   Svelto, O., Principles of Lasers. New York: Plenum Press, 1976.

TAK65   Takayanagi, K., Adv. At. Mol. Phys. 1, 149 (1965).

TOE68   Toennies, J. P., in HAR68, p. 157.

TOE74   —— in EYR74, p. 228.

TOE74a   ——, Chem. Soc. Rev. 3, 407 (1974).

TOE76   ——, Ann. Rev. Phys. Chem. 27, 225 (1976).

TRO75   Troe, J., in MIL76, Part B, p. 835.

TRU76   Truhlar, D. G., and R. E. Wyatt, *Ann. Rev. Phys. Chem.* **27**, 1 (1976).

TRU77   —— and ——, *Adv. Chem. Phys.* **36**, 141 (1977).

TRU79   —— and J. T. Muckerman in BER79, p. 505.

TRU79a —— and D. A. Dixon in BER79, p. 597.

TRU81   ——, ed., *Potential Energy Surfaces and Dynamics Calculations.* New York: Plenum, 1981.

TUL76   Tully, J., in MIL76, Part B, p. 217.

TUL77   —— in BRO77, p. 206.

TUL77a —— in *Semiempirical Methods of Electronic Structure Calculation,* Part A (G. A. Segal, ed.), p. 173. New York: Plenum, 1977.

TUR78   Turro, N. J., *Modern Molecular Photochemistry.* San Francisco: Benjamin/Cummings, 1978.

VAN73   Van den Bergh, H. E., M. Faubel, and J. P. Toennies in FAR73, p. 203.

WAL76   Walther, H., ed., *Laser Spectroscopy of Atoms and Molecules.* Berlin: Springer, 1976.

WAL79   Walther, J., and K. W. Rothe, eds., *Laser Spectroscopy IV,* Springer Series in Optical Sciences, Vol. 21. Berlin: Springer, 1979.

WAL81   Wallace, S. C., in JOR81, Part 2, p. 153.

WEI79   Weiner, J., in GLO79, p. 587.

WEI81   Weitz, E., and G. W. Flynn in JOR81, Part 2, p. 185.

WIL74   Wilson, L., ed., A.F.O.S.R. Workshop Report, *State-to-State Molecular Dynamics,* 1974.

WIL77   ——, S. N. Suchard, and J. I. Steinfeld, eds., *Electronic Transition Lasers II.* Cambridge, Mass.: MIT Press, 1977.

WOL70   Wolfgang, R., *Acct. Chem. Res.* **3**, 48 (1970).

WON73   Wong, Y. C., and Y. T. Lee in FAR73, p. 383.

WOO81   Woodin, R. L., and A. Kaldor in JOR81, Part 2, p. 3.

WYA77   Wyatt, R. E., in BRO77, p. 185.

WYA79   —— in BER79, p. 567.

YAR75   Yariv, A., *Quantum Electronics,* 2nd ed. New York: Wiley, 1975.

ZAR74   Zare, R. N., and P. J. Dagdigian, *Science* **185**, 739 (1974).

ZAR79   —— in FAR79, p. 7.

ZAR80   —— and R. B. Bernstein in PHY80, p. 43.

ZEW78   Zewail, A. H., ed. *Advances in Laser Chemistry,* Springer Series in Chemistry and Physics, Vol. 3. Berlin: Springer-Verlag, 1978.

ZEW80   —— in PHY80, p. 26.

ZOR65   Zorn, J., ed., *Experiments with Molecular Beams, Selected Reprints,* Vol. 1. New York: American Institute of Physics, 1965.

Index

Subjects

DATE DUE